Map Data Processing

Academic Press Rapid Manuscript Reproduction

Proceedings of a NATO Advanced Study Institute
on Map Data Processing
held in Maratea, Italy, June 18–29, 1979

MAP DATA PROCESSING

Edited by

HERBERT FREEMAN
Rensselaer Polytechnic Institute
Troy, New York

GOFFREDO G. PIERONI
Dipartimento di Matematica
Università della Calabria
Roges de Rende
Cosenza, Italy

1980
ACADEMIC PRESS

A Subsidiary of Harcourt Brace Jovanovich, Publishers
New York London Toronto Sydney San Francisco

ACADEMIC PRESS, INC.
111 Fifth Avenue, New York, New York 10003

United Kingdom Edition published by
ACADEMIC PRESS, INC. (LONDON) LTD.
24/28 Oval Road, London NW1 7DX

Library of Congress Cataloging in Publication Data

Nato Advanced Study Institute, Maratea, Italy, 1979.
Map data processing.

Proceedings of a NATO Advanced Study Institute held June 18–29, 1979, sponsored by NATO scientific Affairs Division.

Includes index.
1. Cartography—Data processing—Congresses.
I. Freeman, Herbert. II. Pieroni, Goffredo.
III. North Atlantic Treaty Organization. Division of Scientific Affairs. IV. Title.
GA102.4.E4N37 1979 526′.028′54 80-23015
ISBN 0-12-267180-5

PRINTED IN THE UNITED STATES OF AMERICA

80 81 82 83 9 8 7 6 5 4 3 2 1

CONTENTS

CONTRIBUTORS

Numbers in parentheses indicate the pages on which authors' contributions begin.

Richard Baxter (223), Building Research Station, Department of Environment, Garston, Herts, England

A. Raymond Boyle (27), Department of Electrical Engineering, University of Saskatchewan, Saskatoon, Saskatchewan, Canada

M. F. Costabile (309), Dipartimento di Matematica, Università della Calabria, Roges di Rende, Cosenza, Italy

Larry S. Davis (169), Computer Sciences Department, University of Texas, Austin, Texas

Avi Degani (207), Department of Geography, Tel Aviv University, Tel Aviv, Israel

Silvano Di Zenzo (247), IBM-Italia, Rome, Italy

Herbert Freeman (151), Rensselaer Polytechnic Institute, Troy, New York

Michael F. Goodchild (191), Department of Geography, University of Western Ontario, London, Canada

C. Guerra (309), Dipartimento di Matematica, Università della Calabria, Roges di Rende, Cosenza, Italy

Robert M. Haralick (63), Department of Electrical Engineering, Virginia Polytechnic Institute and State University, Blacksburg, Virginia

Ellen M. Knapp (47), Geoscience Systems Department, Computer Sciences Corporation, Silver Spring, Maryland

S. Levialdi (279), Laboratorio di Cibernetica, CNR, Arco Felice, Napoli, Italy

George Nagy (119), Department of Computer Science, University of Nebraska-Lincoln, Lincoln, Nebraska

G. G. Pieroni (309), Dipartimento di Matematica, Università della Calabria, Roges di Rende, Cosenza, Italy

Azriel Rosenfeld (137), Vision Laboratory, University of Maryland, College Park, Maryland

Bruce Schachter (355), Ground Systems Department, General Electric Company, Daytona Beach, Florida

Linda G. Shapiro (101), Department of Computer Science, Virginia Polytechnic Institute and State University, Blacksburg, Virginia

Dieter Steiner (1), Department of Geography, Swiss Federal Institute of Technology (ETH), Zurich, Switzerland

Helmut Werner (331), Institut für Numerische und Instrumentelle Mathematik und Rechenzentrum, Westfälische Wilhelms-Universität, Münster, Westfalen, Federal Republic of Germany

Joan S. Weszka (265), IBM Corporation, Austin, Texas

PREFACE

A NATO Advanced Study Institute was held in Maratea (Potenza), Italy, during the period 18–29 June 1979 on the subject of "Map Data Processing." This book contains a collection of most of the papers presented at that institute. The institute was organized to bring together the leading specialists in the field of map data processing for exchanging ideas and for setting directions for future research and development. An important secondary objective was the bridging of some of the still-existing gaps of communication between researchers in the geography-related disciplines and those in the pattern-recognition , image-processing-, and computer-related disciplines. A total of 77 persons participated in the institute. They came from 16 different countries and were drawn almost equally from the two discipline groups.

The topics addressed by the institute were hardware and software systems for map data processing, spatial data structures, data encoding, map data analysis and manipulation, pattern recognition and image processing, approximation techniques, computerized cartography, and specialized map data processing applications. The book should be a valuable reference source to all persons working in the fields of geography, cartography, computer image processing, computer graphics, and remote sensing.

The support of the NATO Scientific Affairs Division in sponsoring this Advanced Study Institute is gratefully acknowledged.

A MINICOMPUTER-BASED GEOGRAPHICAL DATA PROCESSING SYSTEM[1]

Dieter Steiner

Department of Geography
Swiss Federal Institute of Technology (ETH)
Zurich, Switzerland

An interactive digital system for the processing of geographical data is described. It consists of a minicomputer, a color display unit and a map digitizer and has offline access to a drum scanner/recorder. It can handle data from various sources (image data, map data, statistical data) and it uses different data formats (raster, polygon and serial) side by side with some possibilities of converting from one format type to another. An image display and manipulation software system that operates on both raster and polygon data is completed and operational. A statistical software system that accepts serial data as input and produces output in graphical and map form is under development. Beyond its application for academic teaching and research needs the processing system as presented here is believed to have a potential as a possible tool for local and regional planning units. It is seen as a flexible and highly user-oriented alternative to large centralized geographical information systems.

1. INTRODUCTION

This paper describes an interactive digital system for geographical data processing. Its major hardware components are a minicomputer, a color display unit and a map digitizer. It also includes offline access to a drum scanner/recorder combination. A software system has been developed over the last three years.

Originally, the idea was simply to have a display facility for the handling of digital image data obtained from remote

---

[1]The software for this system has been developed with partial support from the Swiss National Foundation for the Advancement of Scientific Research (grant 2.601-0.76), "Quantitative Analysis of Multidimensional Remote Sensing Data" .

ISBN 0-12-267180-5

sensors such as Landsat. It has broadened since to a concept of a much more general geographical data processing system. Obviously, it is in the first place geared to the Department of Geography's own research and teaching needs. Actually, we put a major emphasis on this latter application. We feel that teaching efficiency is increased considerably by the possibility to work with students on an interactive system and discuss the results on the spot. Beyond this academic use, however, we believe that a minicomputer-based geographical processing system such as the one described here may very well - with a few possible alterations and additions - become an attractive and low-cost tool for local and regional planning units. We see this as an alternative to large centralized geographical data banks and information systems. There is widespread dissatisfaction with such systems for the following reasons:

(1) With increasing size of an information system the purely computer-technical problems begin to dominate, a fact which hampers the discussion and understanding between computer scientists and users. There is the danger that large systems will develop according to their own laws which may be far removed from practical application needs.

(2) As far as application considerations do, in fact, enter the development of large systems, they must reflect the most commonly occurring concerns. Only then can such a system be economical and serve a user community of maximum size. It is equally clear that it will not be capable of meeting many less common and specific needs of individual users.

(3) Maps generated on a line printer or a plotter are standard products obtained from the usual type of geographical information system. To experiment with different variations of data representation is hardly possible. Except for the more special map reproduction and image processing systems that make use of graphic and raster displays, resp., user-controlled display facilities that would allow experimental flexibility have not been a common feature of large information systems.

(4) The possibility for quantitative data analysis are usually rather restricted. It is desirable that a geographical

data processing system contains routines for the more common types of quantitative methods and models.

(5) Some information systems offer interactive possibilities in the form of a query language. For a user, however, the learning of such a language is only worth the trouble if he is in contact with the system regularly. A general hindrance here is that the user may have had no influence on the development of the system, be physically removed from it and have the general feeling that he is communicating with a black box.

(6) Existing information systems are often one-sided in that they are based exclusively on one type of data (such as land cover data only, or census data only) and make use of one type of data format only (either polygon format or raster format).

In the more recent past, further improvements and cost reductions in the field of small computer technology have led to a rapid decentralisation of computing power with, at the same time, a trend towards computer networks. We believe that geographical information systems will also follow this development, which opens up new possibilities for them. Let us point out especially the following aspects:

(1) Systems with a limited regional coverage contain relatively small data sets. This eliminates or at least lessens the necessity of having the data organized in a data bank with a complicated structure.

(2) Small systems are more flexible in that they are less dependent on observing restrictive norms for the sake of efficiency. In particular, there is an increasing possibility to integrate different data sources and to employ different data formats side by side.

(3) A small system is under the direct control of a user and, consequently, can be fully geared to the specific needs of that user. It can also be considered that needs change and that new needs will arise with the availability of a dedicated system. This is made possible through a modular development.

(4) The more simple methods from the highly developed large systems and basic statistical software packages can be used as building blocks. For the handling of problems that

exceed the capacity of the small system (more complicated calculations, larger data volumes) one can always transfer the data to a larger computer, possibly via a direct link. To facilitate such data transfer the establishment of machine independent data formats should have high priority.

(5) By using a raster display unit the already established methods of image processing can be generally applied to the handling of geographical data. The optimum exploitation of the possibilities offered by presently available display hardware will reduce the need for carrying out calculations on the data. Also, spatial information can be represented on the display monitor readily at any time and one can find the variant of that representation best suited for visual evaluation.

It is exactly along these lines that the system presented in this contribution has been developed. It is capable of doing operations in any of the three major areas:

(1) Display and processing of remote sensing image data in digital form (for example, digital scanner data, or digitized air photos).

(2) Display and processing of digitized map data in either the raster or the polygon mode.

(3) Analysis of statistical data which are - in contrast to the data just mentioned above - not directly location-specific, display of results in graphical form (frequency distributions, regression lines, trend curves, etc.) and, if applicable, in map form by using a geographical base as a reference.

As an overview, Fig.1 shows typical data flows from these three main data sources through input conversions to data storage in particular formats to output conversions. Indicated are also major cross-links. In accordance with points 4 and 5 above the system contains its own software for display operations and for doing a range of calculations on small data sets. Larger data volumes are handled by a combination of interactive work on the minicomputer facility and batch processing on a large computer. At any rate, it is obvious that the display unit plays a central part and that its utility is greatly enhanced by the fact that it is under interactive user control.

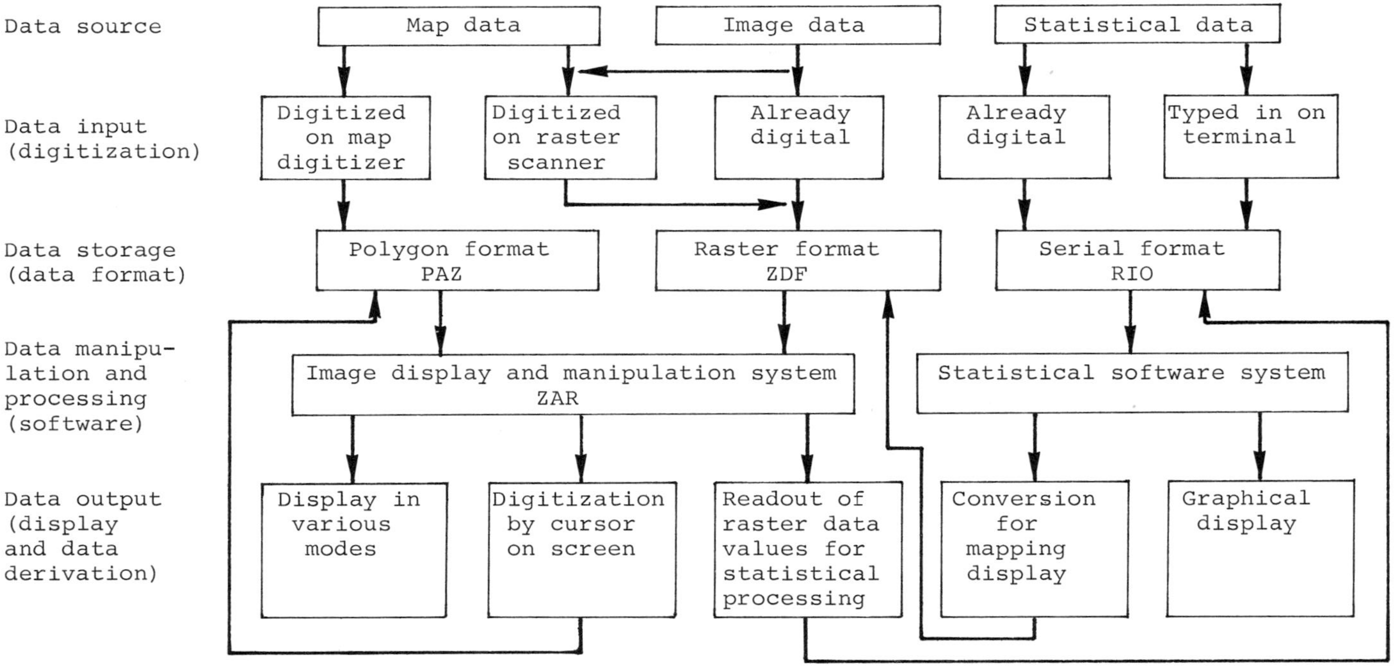

FIGURE 1. Structure of the geographical data processing system at the ETH Zurich with typical flows from data source to data output.

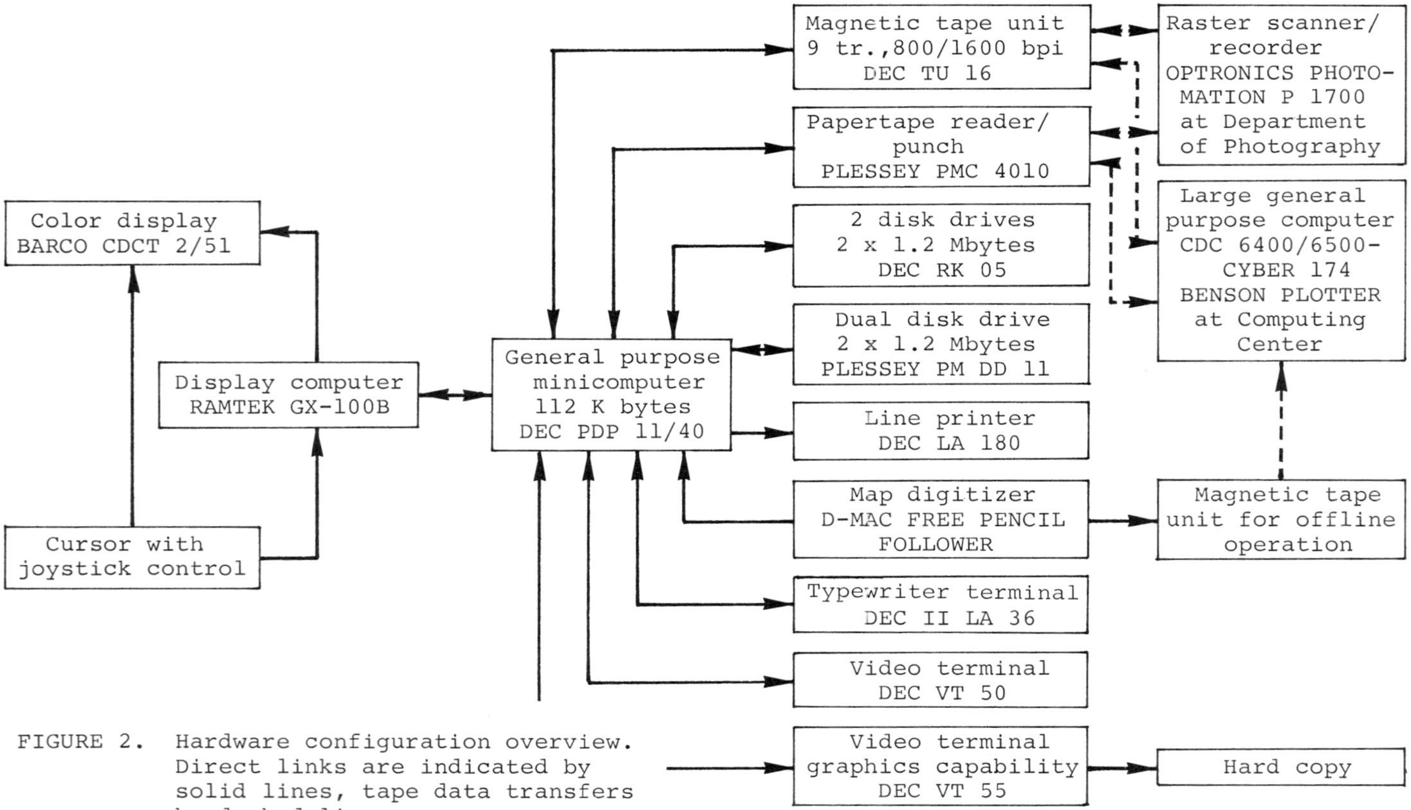

FIGURE 2. Hardware configuration overview. Direct links are indicated by solid lines, tape data transfers by dashed lines.

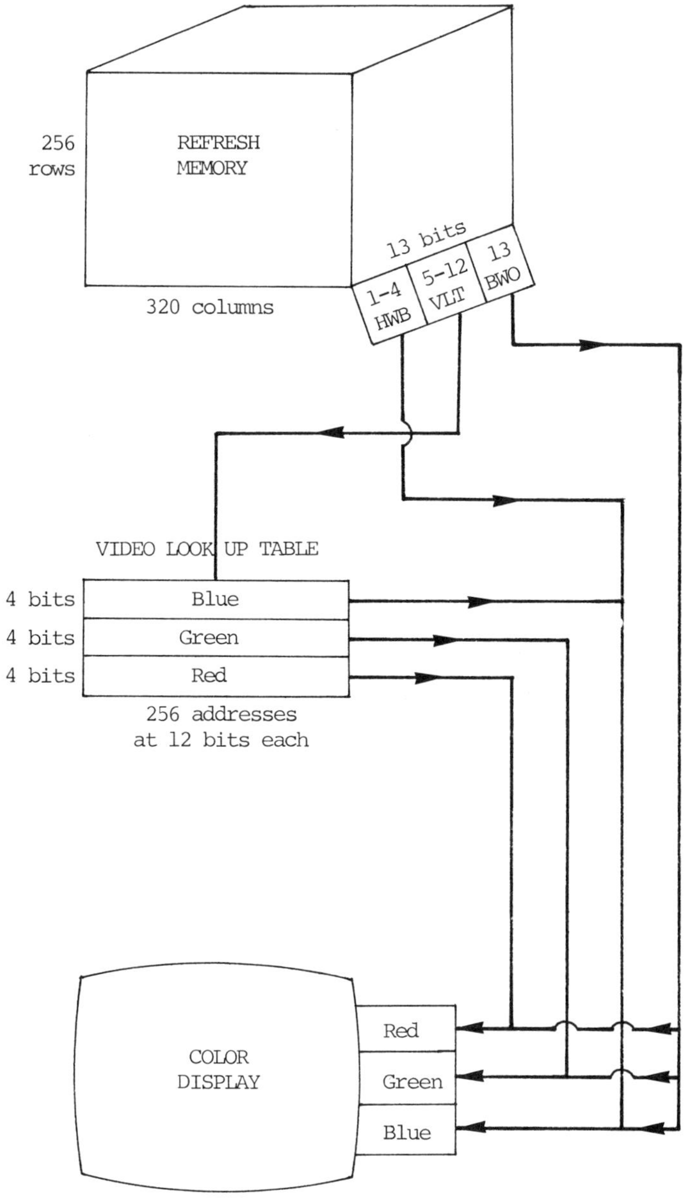

FIGURE 3. Ramtek display system with refresh memory (channels: HWB = Hardwired Blue, VLT = Video Lookup Table, BWO = Black-and-White Overlay), video lookup table and color monitor.

2. HARDWARE

Our system is based on a general purpose minicomputer of the DEC PDP 11/40 type and a special purpose Ramtek GX-100B computer which supports the display facility. The present hardware configuration is depicted in detail in Fig. 2. Also indicated are data transfer possibilities by tape to and from the Institute's Computing Center and the Department of Photography's drum scanner/recorder system.

A direct data link to the Computing Center as well as data transfers to other computing facilities (such as the Department of Cartography's computer-assisted mapping system with a high-quality flat bed plotter or the Center for Interactive Computing with a graphics system) may be considered in the future. As possible hardware additions we envisage the purchase of a matrix printer/plotter and a large disk drive.

The Ramtek system's refresh memory provides a raster display in color with 256 rows and 320 columns and with 13 bits per pixel. These are divided up into three system channels as follows (see Fig.3):

(1) Hardwired Blue (Bits 1 to 4): The first 4 bits go directly to the blue converter of the color monitor. No manipulation is possible with respect to the translation of data values to display intensities except through previous software control.

(2) Video Lookup Table (Bits 5 to 12): The next 8 bits go to a video lookup table with 256 different addresses (corresponding to the 8 bits of input) and with a 12-bit output from each address. This can activate all three converters (blue, green and red) of the monitor with 4 bits each. The video lookup facility provides a means for translating the input data into arbitrary display intensities. This translation can be achieved by online operations or through the use of prestored tables.

(3) Black-and-White Overlay (Bit 13): This activates all three converters equally. It is used for alphanumeric and graphical overlay purposes (representation of numbers, letters,

points and lines). Graphical data given in the form of rectangular coordinates are converted to the raster format by hardware. It means that the Ramtek system can be used for graphical output as well. However, due to the relative coarseness of the raster, there are obviously limitations; a slant line will have a very distinct step-like appearance.

The reason for the existence of the special hardwired blue channel is the fact that at the time the system was purchased only a hardware configuration with an 8-bit input and a 12-bit output video lookup table was available. When working with one-channel data there is no problem. The table can be entered with 256 different levels which are then translated to a 16 level one-, two or three-color representation. However, 8 bits of data are not suited to three-channel data (for example multispectral scanner data) if one wants to have 4 bits per channel and, correspondingly, per color. Consequently, it was decided to feed one channel of data (4 bits) directly to the blue converter. The two other channels with 4 bits each are then routed via the video lookup table and normally used to drive the green and red guns of the tube only.

A problem is the permanent recording of information as it appears on the color display. For simple graphical representations the line printer can be used for this purpose. For data in image form, however, this is not very practical. For permanent documentation the screen can be photographed directly. For higher quality recordings which also provide the possibility of multiple copying we have to resort to the Department of Photography's drum recorder. This is associated with two major problems. First, operations done directly by the display computer have to be duplicated first by software to bring the data into a form which is identical to the display. Second, the preparation of color material is a somewhat tedious job in that the drum recorder can only produce black-and-white output; color separations have to be prepared and combined to a composite by photographic means.

3. SYSTEM SOFTWARE

The PDP computer uses DEC's RSX-11M operating system which can be characterized as a multitasking, multiuser, real timing and interactive system. It supplies the user with a Monitor Control Routine (MCR), a collection of commands for operating the installation and for calling various utilities. These utilities comprise the usual service programs for file transfer and editing, Fortran compilation, assembly language compilation, task building and library maintenance.

For the Ramtek computer no system software was available from the manufacturer. The driver as well as all basic display functions had to be written internally. Similarly, a driver and the necessary support programs were developed for the online operation of the D-MAC map digitizer.

4. DATA FORMATS

According to the kinds of data occurring with the variety of tasks for which the system has been designed (see 1) three different types of data formats have been developed and are now in use: Raster format, polygon format and serial format. We also note that some conversions from one format to another are possible, in particular:

(1) Polygon to raster: An area boundary given as a vector sequence is overlaid on a raster; the raster cells belonging to that area are determined.

(2) Raster to serial: The values of raster cells contained within a specified area are read out and stored as a data series.

(3) Serial to raster: Spatial statistical data are combined with a geographical base and converted to a raster file (map display).

(4) Raster to polygon: The boundary of an area in a raster file is extracted and stored in the form of a vector sequence.

4.1 Raster Format (ZDF)

This format is used for any type of data for which the sequence of values has an implicit locational connotation. Typical examples are multispectral data from aircraft scanners or Landsat, map information digitized on a raster scanner and census data converted to a raster map representation.

The ZDF format has been designed as a machine-independent format to facilitate data exchange between different institutions. Machine independence is achieved by the following two properties:

(1) Conventions regarding value coding such as standardized coding of reals and fixing the number of bits for integers.

(2) Availability of all information necessary for access and processing (e.g., geometric origin, channel designations, etc.) in the data file itself in the form of a header record, channel records and annotation records (1).

The first physical data record for a scan line (or more generally a row of data) contains at the beginning locational information (number of the scan line, relative shift of present scan line to the previous one if any). The data itself are recorded in pixel-interleaving fashion, i.e., all channel values for the first pixel are given in sequence, followed by all channel values for the second pixel, etc. This storage mode is particularly suitable if operations are to be carried out pixel by pixel on several or all channel values. It has its drawbacks if only one channel of data is wanted.

The data values can comprise any number of bits from 1 to 32. They are stored in compact form and, consequently, must be packed for write and unpacked for read operations, resp. The physical structure of a ZDF file provides direct access to individual rows of data. This means that data correction or updating can be simply achieved by calling up one row at a time and replacing the values as needed. The direct access feature also applies to the reading of data from tape, i.e., a raster file on tape can be read as if it were on a random access device. Conversely, a file cannot be written directly on tape.

4.2 Polygon Format (PAZ)

This is a format for graphical data such as points, lines and closed polygons. Examples are: for points: pass points; for lines: street segments; for closed polygons: area boundaries. Generally, we use the term polygon for a sequence of points whose location is defined explicitly by x,y-coordinates and which may or may not be connected. PAZ data are commonly created by digitizing map information on the digitizer table or by interactive digitization of some features in a raster image displayed on the color monitor.

Two features of PAZ are particularly notable. First, similar to the ZDF, it is a machine-independent format. Relevant information is provided in a header record and in a system index record (containing for each polygon the number of the starting record and the length of the polygon). Second, it has been designed in such a way that any number of thematic data values can be added to each pair of coordinates. Consequently, it is possible, for example, to store field observations recorded at particular points in this format rather than in the inverse fashion given by the RIO format to be discussed below. The data records contain all values (x- and y-coordinates plus possible further values) for the first point in sequence, then all values for the next point in the same polygon, and so on. Direct access is available per polygon.

An additional feature of the PAZ format is that each file is provided with a user index whose structure can be determined by the user. A fixed structure is being employed in connection with the ZAR image display and manipulation system (see 5.2). Labels on three different hierarchical levels are assigned to individual polygons so that they can be called up selectively. This structure can be understood as a table in normal form of the first order (2).

4.3 Serial Format (RIO)

Serial data are linear arrays of values of one variable or attribute observed for a number of different objects. In German such a series is called a "Messreihe". Examples are the number of inhabitants for all administrative units of a certain administrative level of a country (these are data that are serial in the spatial sense) and the average temperature for all months during a 30-year period for a given climatic station (these are data that are serial in time). If the data are serial in spatial terms they do not, however, carry direct (implicit or explicit) locational information. Spatial location may be inferred indirectly through an external index. Such an index may, for example, associate the sequence numbers of data values with the names of administrative units, which in turn can be found on a map. If desired, it would, of course, be possible to store x- and y-coordinates as two additional series for the same spatial units.

The RIO format is designed for data storage on disk and for easy file maintenance (changing or updating the contents of a file). Each series can be accessed individually. A table of content is maintained for each file. It gives for each series its name and number, the number of the first record and the length of the series (number of records). The format is particularly suited for any kind of statistical processing. It is not, at present, machine independent. Further improvements are currently being implemented together with a system of statistical routines (see 5.3).

5. USER SOFTWARE

5.1 Input/Output Routines

For each type of data format (ZDF, PAZ and RIO, see 4) the necessary input and output routines for creating new files, updating and reading existing files exist. Moreover, the RIO package provides the possibility of variable transgeneration, i.e., new derived series can be produced, for example, by combining two or more existing series into a new one, or by calculating sums, differences or ratios for corresponding values of existing series.

5.2 Image Display and Manipulation System (ZAR)

A major development during the last two years has been the establishment of an interactive image display and manipulation system that is completed in its basic form but is designed in such a way that it can be expanded at any time (3).

ZAR is not an image processing system in the actual sense, i.e., it keeps actual computational operations on images to a minimum. Instead, it makes optimum use of the existing hardware facilities for interactive display and manipulation purposes. It presents to the user a number of basic functions with different options. A particularly important one among them is a routine that provides the possibility of manipulating images through changing the contents of the video lookup table. The underlying rationale is to give the user a means for optimum visualization of pictorial information on the basis of which he then can take decisions regarding further actions. Such an action might consist of interactively defining parts of an image data file for which a certain operation should be carried out on the Computing Center's large computer.

ZAR works with location-specific data of both the raster (ZDF) and the polygon (PAZ) type. It is genuinely interactive in the sense that its callable functions are available in parallel, i.e., the user can activate them in any sequence (with some notable exceptions: some functions are blocked during the time another function is in use). In short, the machine reacts to the user. This is in contrast to a pseudo-interactive system whose use differs from normal batch processing only in that parameters can be defined by the user at execution time; the sequence of operations is fixed, however. In other words, the user has to react to the machine.

Most ZAR functions comprise a number of options between which the user can choose. However, these options must be used

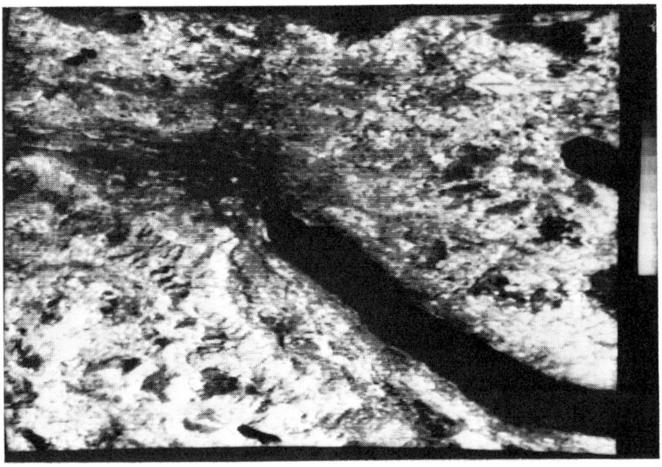

FIGURE 4. Section of a Landsat image, channel 7 (infrared), showing the city of Zurich and sourroundings. The data were read in using the ZAR functions DIM(R), option a (single channel display) and DPL. A normal gray scale (16 levels) was generated by means of the function VLT, option a (interpolation).

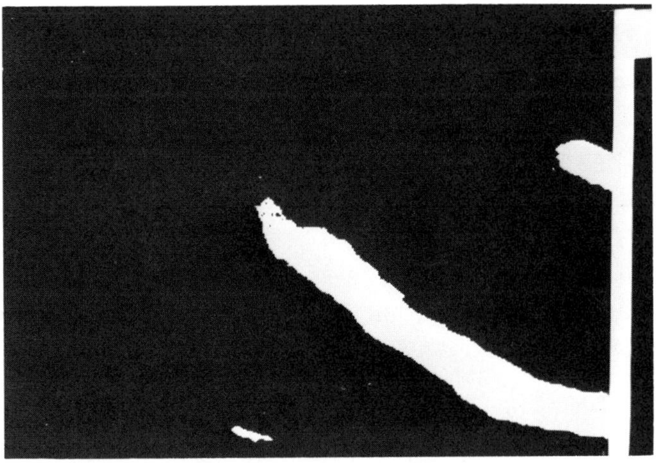

FIGURE 5. Same image as in Figure 4. A level slice (comprising all intensities from 0 to 27) achieved with ZAR function WND depicts water areas (white).

normally in a particular sequence so that a function can be carried out and terminated properly. According to the definition just presented above then the use of options within a function is pseudo-interactive. The following is a brief description of some of the ZAR functions. Some illustrative examples are given in Figures 4 to 11.

(1) FPS (File Portion Selection): Displays the number of the first and of the last row and of the first and the last column as well as row and column increments for the file in question. The user selects the portion of the file to be displayed by specifying a coordinate range in terms of rows and columns. Also, a magnification factor can be defined.

(2) DIM(R) (Display Mode for Raster Data): The user has the following options:

a) One-channel display (can be used, for example, for subsequent level slicing).

b) Two- to three-channel display for subsequent color composition. For both this and the previous mode the data will be transferred from the file to the refresh memory in unaltered form.

c) Intensity transformation: The user is presented with a two-dimensional diagram with the horizontal axis as input scale and the vertical axis as output scale. By using the cursor the user defines a piece-wise linear function that relates output to input (see example in Figure 6). This can be done for one or more channels. The input values will be transformed accordingly before being transferred to the refresh memory.

d) Parallelpiped classification: The user defines a range of values for one or more categories of features and for one or more channels. The pixels which meet the given criteria obtain a numerical classification code. These codes will be loaded into the refresh memory.

(3) DIM(V) (Display Mode for Vector Data): The user defines a selection of polygons to be displayed via the three-level hierarchy provided by the user index (see 4.2) and chooses one of the following display options:

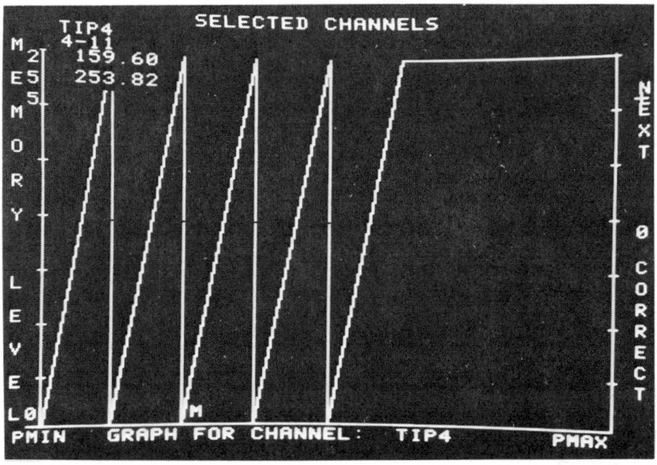

FIGURE 6. Diagram for an intensity transformation defined by means of the ZAR function DIM(R), option c. The transformation results in a multi-cycle contrast enhancement.

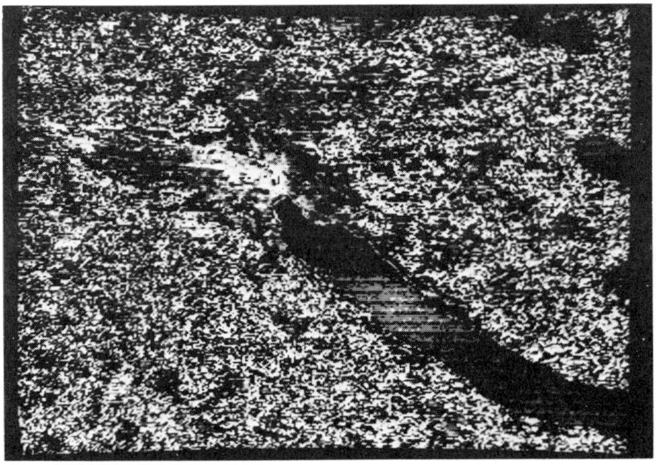

FIGURE 7. Same image as in Figure 4, but read in using the ZAR functions DIM(R), option c (with transformation as defined in Figure 6) and DPL. The resulting picture shows increased local contrasts. However, the noise is enhanced too.

a) Pixel: Each point is shown separately.

b) Vector: Sequential points are connected by lines.

c) Polygon: As in b), but, in addition, the last point is connected with the first one.

(4) DPL (Display): Reads the designated file from the storage device and transfers the data to the refresh memory in the mode selected by DIM.

(5) VLT (Video Lookup Table): Fills the video lookup table with values defined by the user. Options are:

a) Interpolation: The user defines an address range (corresponding to a minimum and a maximum value on the input intensity scale) and parallel to it an output intensity range. The assignment of output levels to input values is achieved through linear interpolation. This option is used for regular one-channel black-and-white and three-channel color composite displays.

b) Color mixing: The user creates any color desired (this color is displayed) and assigns it to a selected address range.

c) Loading: The user calls up an already existing table in file form and loads it into the video lookup table.

(6) WND (Windowing): Selects one of a number of prefabricated color scales (can be a single constant color). The chosen scale is filled dynamically into the video lookup table, i.e., it can be stretched or compressed in a harmonica-like fashion by designating a lower and an upper boundary on the input intensity scale. Once an interval has been fixed (for example, it is 10 intensity levels wide), it can be moved up and down on the input scale. This operation achieves level slicing.

(7) DIG (Digitization): Creates a PAZ file by selective coordinate digitization of a raster image presented on the color screen under cursor control. The purpose might be to define the outlines of certain areas within the image. The result of the operation is displayed instantaneously in one of several possible modes (similar to the modes available with DIM(V)). The geometry of the PAZ file (row and column coordinates) is taken over from the ZDF file.

ZAR also provides the possibility to read out the values of pixels contained within a closed polygon defined by the DIG function and store them as a RIO series for statistical processing.

The ZAR system software cannot be applied only to image data but it can be used equally well for the handling of digitized map data and raster map display of statistical data. For example, raster-scanned color separation sheets of a topographic map can be combined into one file and treated as different channels. Likewise, different statistical attributes of administrative areas such as census variables can play the role of channels when RIO data are converted to a raster map display (see 5.3).

The ZAR system is presently being used for the following projects:

(1) Establishment of a digital terrain model for the test-site Grindelwald (Bernese Oberland) of the Swiss MAB project. Raster-scanned contour line transparencies provide the starting point.

(2) Geometrical fitting of multispectral scanner data to the Swiss Planning Information System (which is based on 100m x 100m raster cells) for the purpose of land use updating.

(3) Preparation of raster data (such as those in the Planning Information system) for quantitative analysis by overlaying boundary networks digitized on the map digitizer and converting them to RIO files.

In each case the PDP-Ramtek system is used mainly for interactive operations while the more involved calculations are done on the Computing Center's large computer.

5.3 Statistical Software System

A large number of main programs and subroutines for doing various types of statistical analysis (such as regression, factor, discriminant and cluster analysis; markov chains, matrix methods for population projection etc.) already exist. They work, however, on ASCII data files and are presently being converted for direct access to RIO files. At the same time they are being assembled into a modular system with a structure similar to ZAR.

On the output side the statistical software system contains routines for graphical displays on the color monitor (for example, histograms, scattergrams and function plots). As an alternative, corresponding black-and-white output can also be produced on the VT 55 video terminal, which has a (limited) graphics capability, or on the line printer.

Another output feature which must be present in a statistical software system operating on geographical data is the display of spatial serial data (such as census data) in map form. The first version of a program package (not yet integrated into the system) for the production of choropleth maps on the color monitor is available. A geographical base is established for the region to be mapped in the form of a raster (ZDF) file. Each pixel carries a number which identifies it as belonging to a particular administrative unit. One way of producing actual choropleth maps on the display would be to create a new ZDF file in which for each pixel the administrative unit number is replaced by one or more attribute values (single- or multi-channel data). Such a file could then be handled by the ZAR system in the same way as a file of multispectral scanner data. A bottleneck in this approach is the necessity for repeated read operations on the ZDF file if one wants to switch from one attribute (or attribute combination) to another; reading a raster file is a fairly slow process.

Consequently, a more attractive solution has been implemented in the present program package. It works as follows: The base file (containing the administrative unit numbers) is transferred to the display unit's refresh memory once. The numbers in this file now constitute addresses for the video lookup table facility. For each attribute a separate table can be constructed and stored as a VLT file by doing the following: Divide the range of attribute values into a number of classes, assign each administrative unit to the appropriate class, select a suitable color scale and determine the resulting color combination for each class. The content of the video lookup

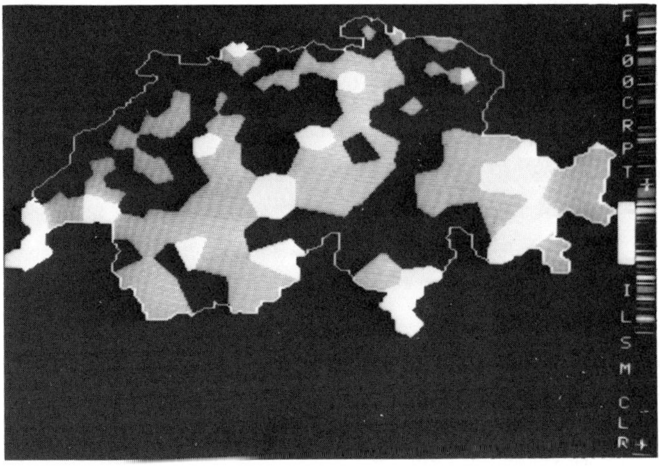

FIGURE 8. Example of a proximal choropleth map with the districts of Switzerland (crude version without lakes). It shows percent employment in the tertiary sector in three classes. The map was produced by using the district numbers as VLT addresses and by loading a prestored VLT file (ZAR function VLT, option c).

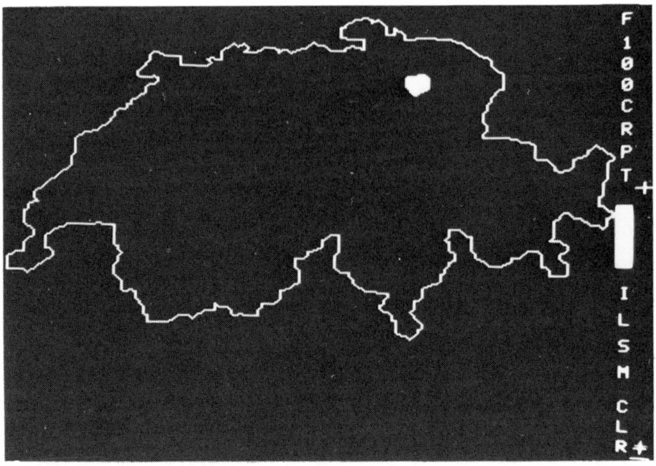

FIGURE 9. District map of Switzerland. By using the ZAR function WND a single district (in this case no.100) can be located (shown in white).

table can be changed instantaneously. Consequently, the rapid switching from one attribute to another becomes possible by simply loading the appropriate VLT files in turn (see figures 8 and 9). It is also perfectly possible to create VLT files for direct two- or three-attribute comparisons. A particularly nice possibility is the generation of "correlation" maps: One attribute is shown in one color, a second one in the color that is complementary to the first one. If there is perfect agreement between the two attributes in terms of class memberships of spatial units the map will appear in black-and-white. Any deviations from such an agreement will show up in color.

5.4 Separate Program Packages

A number of separate program packages are also available on our system which are not integrated in either the image display and manipulation or the statistical software system at present. One is a Fourier transform package which does spatial frequency and directional filtering operations on image data. It has been used for an analysis of Landsat data showing the urban area of Toronto, Canada (4). Another collection of programs exist for the handling of binary areal map data (for example, forested areas or water surfaces that have been obtained by digitizing map color separation transparencies on the raster scanner), using bit quad processing techniques (as discussed in (5). It includes such operations as calculation of areal measures, filtering, pixel labeling, region coloring, boundary extraction (vectorization) and vertex finding (generalization of extracted boundaries) (see figures 10 and 11).

6. CONCLUSION AND OUTLOOK

We have indicated our belief that a small scale geographical data processing system such as the one presented in this paper can become an attractive tool for planning applications. It can be interfaced to a larger computer or a larger information system if the need arises. Locally, however, the user

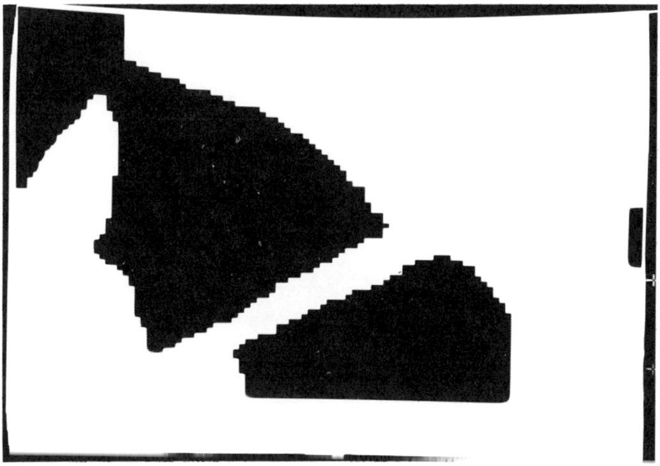

FIGURE 10. Portion of raster-scanned binary map data file showing forested areas (black) in an area west of Zurich. It is displayed with three times magnification (i.e., each pixel is triplicated).

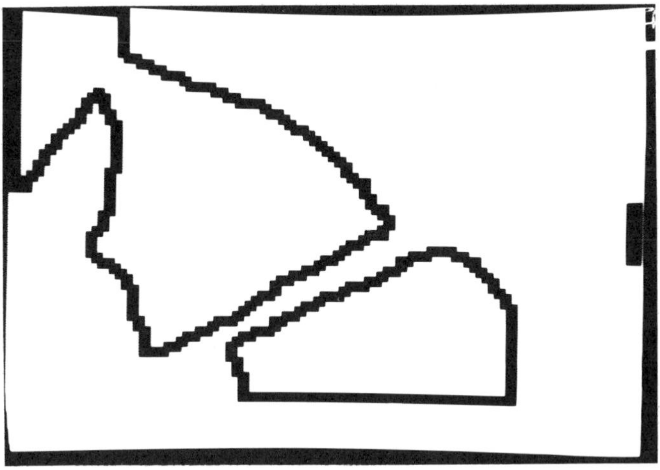

FIGURE 11. Example of an operation performed on a displayed file of labeled pixels (labels denote type of neighborhood) derived from the data shown in Figure 10. By using the ZAR function VLT, option b (color), to select pixels with specified labels, the borders around the forest areas depicted in Figure 10 are obtained.

has full control over the system. A flexible modular development can satisfy increasing and changing demands. Of prime importance is the fact that the interactive use of modern display equipment reduces computational needs in that many decisions can be based upon the visual examination of data displays rather than on the results of some calculations. In this way human and machine capabilities can be optimally combined.

A future project will explore the utility of our computer-display combination in a regional planning context. In particular, we want to investigate the following:

(1) Changes in information content of mapped statistical data with alternative class interval schemes.

(2) The possibilities of using the video lookup table facility for dynamic displays.

(3) The problems involved in representing interaction data.

(4) The integration of image data (obtained by remote sensing) with existing map data by using a common interface (such as a raster based geocoding scheme).

ACKNOWLEDGMENTS

The development of the system described in this paper has been possible only through the efforts of a number of hardworking assistants and doctoral candidates and with some help from other institutions. The following lists the major contributions:

(1) Ramtek display computer, driver and basic software: Richard Benster.

(2) D-MAC map digitizer, driver and basic software: Giorgio Lenz.

(3) ZDF data format, design: Fritz Fasler (Department of Geography, University of Zurich), Otto F. Matt (Swissair Photo and Surveys Company), Anton Paschke, Klaus Seidel (Department of Photography, ETHZ); implementation: Edgar Blum (Department of Photography, ETHZ), Giorgio Lenz.

(4) PAZ data format, design and implementation: Edgar Blum, Giorgio Lenz.

(5) RIO data format, design and implementation: Benedetto Flury, Hans Gilgen, Giorgio Lenz, Anton Paschke.

(6) ZAR image display and manipulation system: Edgar Blum, Hans Gilgen, Giorgio Lenz, Anton Paschke (design only).

(7) Statistical software system: Benedetto Flury.

REFERENCES

1. Seidel, K., "Digitale Bildverarbeitung". Dept. of Photography, ETH Zurich, (1976).
2. Codd, E.F., "Further Normalisation of the Data Base Relational Model". IBM Research Labs., San Jose, CA, (1971).
3. Lenz, G., Gilgen H., Blum, E., Paschke A. and Benster,R., *Int. Archives of Photogrammetry XXII-7*,229, Abt. Luftbildmessung und -interpretation, Univ. Freiburg i/B, F.R. of Germany, (1979).
4. Jefferies, J., "Examination of Digital Fourier Transforms as a Means of Extracting Information about the Physical Form of an Urban Area from Remote Sensing Imagery". M.A. thesis, Dept. of Geography, Univ. of Waterloo, Canada(1979).
5. Gray, S.B., *IEEE Trans. Computers C-20*,551 (1971).

SCAN DIGITIZATION OF CARTOGRAPHIC DATA

A. Raymond Boyle

Department of Electrical Engineering
University of Saskatchewan
Saskatoon, Sask., Canada

The importance of large scale digitization of existing map series is now apparent to most cartographic establishments. Processes of manual and even automatic line following digitization cannot meet the combined needs of data quality, economics and throughput. Scan digitization, on the other hand, has now been shown to be viable, particularly as a result of recent advances in software for line thinning and vectorization. This paper describes one of the methods, together with the data formats used and some of the post-vectorization processes.

I. INTRODUCTION

Coincidently at a number of establishments (CGIS, Canada; CSIR, Pretoria; Mathematical Institute, Bonn; University of Saskatchewan, Canada) experiments have shown that scan-to-line vector conversion of cartographic line data can be achieved at a reasonable cost whereas previous reported work had indicated that it would be extremely expensive, so expensive that, in fact, the process could not be considered for other than essential military work. The reason for this change is still not quite clear, as it has not resulted from a sudden new advance in technology; it would appear that it is perhaps the difference in requirement specifications, as a very slight difference in these can rapidly escalate costs.

ISBN 0-12-267180-5

This change has created a completely new attitude to the 'mass' digitization of existing map series; it has also created a need to answer the new problems that immediately arise. It is the aim of this paper to report on the present state of work in scan digitization, the methods being developed and the next problems to be handled.

For the purpose of this paper it is assumed and accepted that the cartographer needs line data in vector line form and that a raster representation of this is not adequate - that is another point which needs full discussion, however. Moreover, it is also assumed that the mass digitization of existing map series is useful and that for such work colour separation sheets are available with well-defined and clear cartographic line presentation. No attention is given here to the 'possible' automatic digitization of composite paper maps or of unclear or badly drawn working map compilations, although it is hoped to indicate the limits where 'clear' becomes 'unclear'. With automatic digitization of clear lines, the resulting data is of very high quality requiring little interactive editing. This is a major difference other than the one of cost, from manually digitized maps.

The first attention must be given to the previous bottleneck of scan to line vector conversion. All the programs examined by the writer consist of two parts - line thinning followed by line following. Most use look-up tables, often in a relatively simple manner. While the operation on each pixel of an image is fast, it must be remembered that there are approximately 10^8 pixels in one map image using the lowest acceptable resolution. It is therefore essential to reduce the number of passes to a minimum, usually two. Other programs have suffered in timing (and thus cost) by the use throughout of a high level language where assembler routines should have been used. In fact the specification of operational time and cost has been obviously missing from many contracts placed with software houses in the past.

It has now been shown that work can be reduced to two passes of the raster scan data and that the total processing time in a series digital computer should be less than 1 minute on a large main frame and less than 30 minutes on a mini-computer, for a complete separation sheet about 25" square. Actually the minicomputer appears to be longer in time but lower in amortized cost.

In order to obtain efficient conversion certain restrictions should be applied to the input document. It now appears that these are not onerous although some of the older routines such as CGIS are too restrictive for production use.

The conditions applicable to the input document are as follows:

1) The lines should be well-defined. Preferably the original work will have been scribed and this is normal for most map series made in the last twenty years. Copy work from the original must be properly controlled in quality.
2) Line weights should be within a specified range (the U-of-S system for example allows a 4:1 ratio).
3) Gaps between lines should be greater than a specified minimum (usually 1 or 2 pixels).
4) Unwanted data should be absent or removed (usually by opaquing).

5) 'Blob' symbols and very wide lines (greater than 0.4mm (0.016")) can cause difficulty in most systems and are catastrophic in some, e.g. CGIS. A 'blob' symbol may be a small house block as symbolized on a medium scale map.

It should be reiterated that the discussion is only of the digitization of high quality separation sheets as presently made for colour printing.

II. SCANNING

Before examining the scan to line conversion process, it is necessary to understand the scanner operation. Within the last few years, high quality scanners have been made for many applications, particularly for alphanumeric copy work and for the graphic arts industry. At first scanners were used in the analog mode to handle grey scale more easily, but recently digital methods have become more popular. Most use rotating drums and axially moving heads. The better ones have settable resolutions (number of lines per inch) and spot size. Although not essential, these variable facilities are very useful in cartographic work. These drum systems unfortunately tend to be difficult to manufacture and thus become expensive.

It has been reasonably common in graphic arts work to scan an image on one drum and repeat the trace on a scan plotter at the same time. However, the copy that is made can be modified in intensity and contrast, and both of these for selected parts of the image in order to provide enhancement. The same process is used with colour filters on the reading head to produce colour separations for printing purposes.

Many of these scanners can be used for digital scanning of cartographic separations, the signal line being fed to a level discriminator (black/white) before passing to digital logic circuits and a computer storage system.

The drums usually rotate at a constant synchronous speed; 300 rpm is frequently chosen (5 rps) which means that one scan line is output in 0.2 seconds for one complete drum rotation. The timing resolution of this line might be into 10,000 parts over the drum periphery of say 1 meter (40") (the drum must be large enough for the work in hand). This would also infer 10,000 scan lines for 1 meter axial motion resulting in a total scan time of 2,000 seconds or approximately 40 minutes. This scan time is fixed for a certain resolution and dimensions, independently of the amount of information on the separation.

The new breed of flat-bed laser beam scanners appears to be preferable as constructional costs are lower and speeds can be much higher without excessive dynamic difficulties. The time of full scan can easily be reduced to one tenth of that of the drum without running into any major problems of bit data entry or data storage speeds.

The reader will appreciate that a raster scan is one where the separation sheet is automatically examined step by step, each step being the resolution size, normally of about 0.1 mm (0.004"). The output is similar to that used for a black/white television image, but of much greater resolution. Examining a separation of 1 meter x 1 meter at 250 lines/inch will result in 10^8 pixels, which must be stored on an attached unit as they are produced. This unit need not be a computer, but often that is the most convenient device. This amount of data is large but it is well within the speed and bulk capabilities with reasonable planning. The pixel data may be stored as on-off (1,0) bits or in a run length encoded form (i.e., distance from last set pixel).

The optimum resolution in lines per inch (and the related spot size) is always an arguable point; 250 lines per inch is suggested by the writer as reasonable. It means that the spacing of scan lines is 0.1 mm (0.004") each with a set precision of 0.05 mm (0.002"). Thus with a good quality mechanism, a precision of ±0.05 mm (±0.002") could be expected and, as a data representation, this will be as good as could have been originally drawn by hand. There is no concern here with separation stretch or contraction (providing it is an orthogonal change) as registration marks are automatically digitized along with line data and the X and Y scaling is then automatically corrected. The spot size used will be approximately 0.12 mm (0.005") to allow slight overlap between lines. The mechanism must produce evenly spaced parallel straight scan lines at a constant travel speed (normally time in the one axis gives position).

Users who think that they should use a higher resolution must remember that a resolution of 500 lines per inch requires four times the data handling rate and storage, and 1,000 lines per inch requires 16 times. The gains are doubtful when it must be remembered that probably the higher resolution would be passed on to the later scan-to-line conversion program, although of course this is not a necessary corollary of high resolution scanning.

Cartographic requirements are only for black/white levels and this makes the scanner design simpler than if highly linear grey scales must be recorded. The photo detector or detectors are therefore not so critical in output level.

The separation sheets used on a scanner can be black lines on a white base for reflective scanners or negative or positive transparencies for transmission scanners. In practice it is possible to read a transparency on a reflective scanner by mounting a white sheet between it and the base; however, troubles can arise from reflections from the highly polished surface and the double passage of the light through the transparency. The University of Saskatchewan system works with transmitted light and transparencies.

At this time scan times are of the order of one hour using drums. This is uneconomical in amortization cost and throughput. It is expected that times of about 5 to 10 minutes must be used in future work. It is not worthwhile shortening beyond this, as each sheet has to be fitted by hand; this probably requires at least 2-3 minutes even though it is not necessary to position exactly in the rectangular axes.

It should be noted that efficient scan to line vector conversion depends to an appreciable extent on the proper preparation (see section on preparation) and quality of the original document.

III. LINE THINNING

In order for the subsequent automatic line following to be fast and precise, single pixel width lines must be available to it. The data as scanned will typically be 2-3 pixels in width; the width may vary and be as high as 10 pixels or even disappear at points; line thinning must find the center of these variable width lines and record the single pixel data line as output.

Even if vector line data were not to be produced, it would still probably be necessary to carry out the thinning for a raster form of data storage.

Different ways of line thinning have been developed by each establishment interested in this work. There has been more variability here than in any other associated operation. A process widely used has been to nibble away at each side of a line using standard edge detection methods until only one pixel is left. Unfortunately this involves multiple passes on very large amounts of data, resulting in appreciable processing time and thus cost. This method does however automatically (but expensively) find the line weight being handled, as it is inversely proportional to the number of passes. An output of line weight as well as a centre line data can be important to some users.

A method at the opposite extreme to this is the one used at the University of Saskatchewan. It examines each pixel against a standard table of 3 x 3 matrix look-up patterns until one side of the line is found (the furthermost and lowest from the raster line start in actual fact). It does not record this pixel but the successive look-up patterns specify the point as one position back towards the line center (axially or diagonally as the case may be). This results in an exact center line for 2, 3 and 4 pixel width lines. When only one pixel is present step back is inhibited. Thus all lines up to and including 4 pixels in width (0.4 mm (0.016") in the University of Saskatchewan system) have their center lines recorded exactly or as exactly as a digital resolution of 0.1 mm (0.004") allows. Lines greater in width than 0.4 mm (0.016") have usually been removed in the University of Saskatchewan preparation work but, if not, or if small 'blobs' appear on the line or at a junction, then the recorded point is not truly central. This is usually a minor problem as against, for example, the CGIS routine which aborts when lines are greater than 10% above the specified width.

The template matrix used in the University of Saskatchewan program (and at CSIR Pretoria) has two extra pixels to aid in the analysis as they give some prewarning of changes to the progress of the line at any instance and approximate to the advantages of using a 4 x 4 matrix without a time and cost penalty.

At present no line weight specification is given by the U of S program. To do this it is necessary to scan and handle data (at least in the initial part of the analysis) at a much finer resolution than 0.1 mm (0.004") and this can increase costs appreciably.

One difference between programs is in the finding and recording of nodes for junction and ends. In some this is done at the time of line thinning - in others it is done at the stage of line following. It is not believed that this difference is important.

Much more development can usefully be done relative to line thinning, particularly to compare methods for quality of results, time and costs and capabilities. However, at least it is now known that line thinning can be done by a serial digital computer in a reasonable time and at a reasonable cost - even for a dense line separation and on a minicomputer. It is not necessary to use large and expensive parallel processors for this purpose; it may be that small parallel processors and special hardware will eventually be the method used - experiments to date are not conclusive on this matter but are being pursued.

Under the heading of de-symbolization it will be noted that mention is made that some of that work may better be included at the line thinning stage and operate on the line data while it is in raster form. At this time development work in this area is being carried out at the University of Saskatchewan.

IV. LINE FOLLOWING

If vector lines are to be produced, any adjacent pixels (now recorded as 'bits') of any thinned (single pixel) line must be found and re-recorded in an ordered form as a sequence of co-ordinates. In most cases this is done by means of another 3 x 3 matrix look-up pattern. Programs from different sources vary slightly in the exact sequence of operation. The following description is limited to the one used at the University of Saskatchewan.

The scan data is examined line by line until a black pixel is found. The surrounding 3 x 3 pixels are then compared with the standard set of alternative patterns. A decision is made; there may in fact be a number of alternative decisions and in this case the first is used and the others stored for subsequent operations. Decisions will consist of the adjacent pixel to be regarded as the next in the line sequence. The next one now becomes the center of another 3 x 3 comparison and the decision is repeated until there are either no more pixels in that direction (a line end) or so many that a junction is indicated. In either case the tracking operation ceases and a second tracking back down the same line is set in motion. This time there is a difference, however, in that each pixel in turn is recorded as a co-ordinate point along the vector line and then erased from the raster image so that no repetitive operations on it can occur. On its return down the line it does not stop at the original pick-up point, but continues until another line end or junction is found. Operation and recording then cease for that line segment. If there had been any alternative decisions at the starting position (other than the opposite line direction) these are then dealt with in turn until all have been followed and recorded.

The raster line examination then proceeds until another black pixel is found and the whole process repeated. It will be noted that if the first line followed had in fact curved into this raster line once more it would not now be found, as such pixels have already been deleted.

The final result is thus a series of line segments, a segment ending or starting at an end or a junction (the end may also be the end of a working zone, which will be described under de-zoning). At the same time a line directory is developed containing a serial number, start and end co-ordinates, line length and a 'bounding' rectangle; these are valuable data for subsequent operations and save a further pass. Also a node file is constructed indicating differences between floating and junction nodes and zone boundary nodes in X or Y.

The operational speed of this program is high. At first an attempt was made to eliminate the first end search tracking along the line, but as this simplified operation appreciably and only increased time in a minor way, this process has been retained.

V. LINE WEEDING

Weeding is the process of reducing the amount of data representing a line; this usually means reducing the number of absolute or incremental co-ordinates in the line string. It is most simply based on linear segments between the selected points, but could equally be used when these represent points on cubic splines.

On any cartographic line a 'straight' or 'spline' line segment is specified as one that, within the tolerance of the digital resolution used, records the original data exactly. Additional cartographic tolerance might be allowed; but such tolerance depends on the type of cartographic line feature, the map scale used as well as on the subjective opinion of the cartographer.

VI. ZONING AND DE-ZONING

It has been noted that about 10^8 pixels may be produced from a scan and no serial computer can reasonably be expected to handle that number in one operation. It is necessary to 'zone' the data. The size and shape of the zone is a matter of convenience, but most systems operate on about 250,000 pixels at once. This can be equivalent to a zone of 500 x 500 (2" x 2" on the University of Saskatchewan system) or 25 x 10,000 for a long thin strip zone.

Assuming the data is only required to produce line vector output, it is likely that the square zone is most efficient as minimizing the number of contained line segments. However, if the data is to be finally output to a scanner plotter, it could possibly be worthwhile considering the use of strip zones and in fact retaining the vector line data in these zones.

VII. DE-SYMBOLIZATION

Many aspects of cartographic data extraction are included under this heading. The obvious work is concerned with the formulation of continuous data lines from cartographic lines symbolized in different, often discontinuous, manners. Such work can also lead directly to automatic labelling of these data lines.

It is also concerned with the detection of junction situations which can either provide useful information or maybe simply a matter of removal of an unwanted node.

More advanced cartographic analysis may also be done in some cases as, for example, to find where drawn lines split, perhaps indicating the presence of a lake in a river line. The lake outline can be formed as a separate entity joined by the river lines.

Many situations have to be detected and handled in an efficient automatic system. In most cases the operation is carried out on vector line data, although frequently the vector line director itself is adequate. It will be remembered that the line and node directories have been created with de-symbolization in view and contain the following information:

1) line serial #

2) start co-ordinate

3) end co-ordinate

4) line length

In some cases such as double cased roads, it may be preferable to carry out the de-symbolization process on the raster data but this work has not yet been done by the writer.

Considerable aid is given to the de-symbolization routines by the operator's input of the separation sheet type (e.g., drains or roads) and of symbolization specifications.

VIII. DISPLAY AND EDIT STATIONS

At least one display and edit station is an essential part of any 'automatic' digitization system. If the work of digitization is subcontracted then such a station is necessary both at the contractor's and at the originator's premises. A display and edit station is used for inspection and correction of data as well as, most importantly, for labelling.

A display and edit station consists of one or more display screens, a keyboard and/or audio input and a pointer. The screen size itself is often not a matter of major importance, but a high speed of operation of the system is vital. It must be possible to display the selected data from one million co-ordinates (the contents of a densely line single separation sheet) within one or two seconds. It must be possible to point to any line - at any point on that line - and recover requested information (e.g., line serial number, etc.) within one or two seconds. Such high speed operation is difficult to obtain as it involves very efficient programming with an appreciable part in assembler routines, and well designed interfaces. Generally display and edit work is carried out on the vectorized output data (before or after de-zoning) but a useful future addition may be to allow work also on the raster input data. These stations always require local intelligence and storage so that an appreciably dedicated minicomputer becomes mandatory.

When display and edit stations have been used with manually digitized line data, the majority of the work has been concerned with correction, particularly at junction situations. With automatic digitization this type of fault rarely occurs and the main use has moved to interactive labelling. One of the weaknesses of automatic digitization is that it is normally impossible to output lines properly labelled. The human appears to be necessary to do this although the aids indicated under de-symbolization are very helpful.

It has already been stated that operational speed on large amounts of data is essential and many edit stations on the market today are incredibly and unnecessarily slow. Automatic digitization can be fast and economical but these advantages can be lost completely by an inefficient labelling system.

IX. MANUAL DIGITIZATION

The scanning process is far more efficient than manual digitization for lines but unlikely to be so for locational symbology, because of the relatively small numbers of these on most standard map series. The situation could be different in special cases, such as hydrographic charts and botanical maps, but it is only in the aspect of house symbols that an automatic process might usually be considered in topographic mapping.

It takes approximately 2 seconds to locate and point to a location on a digitizer table. However, the operator must also add a label and a keyboard operation tends to be poor, because it usually causes the operator's eyes to deflect from his work, making it difficult to return to the same or an adjacent point. This leads to a high probability of key errors and also of missing locational points. This situation is much improved by the use of audio (spoken word) input via a microphone connected to the computer. A vocabulary of 50-60 words is generally reliable for persons with whom the program has been trained, and this allows simple coding to be carried out with ease, with better reliability than from a keyboard and without causing the operator to lose locational concentration. The situation is even further improved if the system is equipped with audio feedback creating an easily comprehended loop via the operator's ears.

X. AUTOMATIC CORRECTION

It has been indicated that automatic corrections can possibly be made. In almost all cases such operations should be listed as they are done against a reference co-ordinate, so that on checking, rectangles or circles generated by the co-ordinates of this list can indicate to the operator that automatic corrections have already been carried out.

XI. PREPARATION PRIOR TO SCANNING

The first rule in preparation is that all separation copies must be of very high quality as far as line reproduction, sharpness of edge, constancy of line weight and lack of scratches and dust marks. This may sound to be too onerous, but if automatic work free from errors is to be obtained, it is necessary. However, almost as good results can be obtained with a considerable relaxation, although normal quality reproduction standards must still be maintained. (Unfortunately the writer has found that this is not always the case.)

There are four aspects of preparation work:

1) Cartographic inspection

2) Separating out unwanted and wanted data

3) Separating out different line weights

4) Opaquing useless data

XII. FORMATS

At the U. of S. much consideration has been given to the most desirable line data formats, in view of the very large amounts of such data that could be generated rapidly by the scan methods proposed. It is assumed that the data will be stored in vector format as this is more compact than raster format. While optical disk storage systems certainly open the way for efficient storage of digitized map data, nevertheless their advantages must not be lightly squandered.

Not only do we have to consider the most compact storage method, however, but we also have to consider the ease and speed of retrieval, particularly for interactive display and edit, or data 'browsing'. Too complicated a structure causes appreciable time overheads and the advantages of compaction can be lost.

A detailed visual examination of many map sheets showed that within the tolerance of digitization, the line data could usually be represented by a straight line segment sequence of length about 0.02". The curve of line segment lengths showed it to be skewed and, while occasionally segment lengths of only 0.004" were required at 'jagged' points, more frequently longer vectors could be used without loss of detail. How could advantage of this be best taken? Obviously the lengths indicated would depend on factors such as type of separation (coastline or coutour) and the scan resolution used.

First it was considered that all coordinate points along a line should be represented in absolute form. Many cartographic data bases use this method at this time. This would require at least 32 bits for X and Y. The alternatives of Freeman chain (3 bits per coord., after the first) and incrementals (deltas) would both then seem to be advantageous. The lengths of the incrementals can be varied as convenient. In practice, because of normal word lengths and byte addressing, the only alternatives are 4 bits X, 4 bits Y or 8 bits X and 8 bits Y; the use of 16 and 16 has no advantage over absolutes.

It appeared to the writer that 8 bits X and 8 bits Y was most advantageous for data output from the system and cartographic storage. This allowed for vector lengths of up to ±127X and ±127Y elements (or for sharp corners as low as 1X and 0Y) to be entered, so that vectors could be at resolved angles of less than 1°. With this method any irregularity or smoothness of line could be efficiently stored. Even for the long straight lines of urban work it was not too bad as against a series of absolute coordinates representing only line ends, and its use maintained a constant format within the system. The handling of the 16 bit words and 8 bit bytes was obviously advantageous for programming on most minicomputers, and only required a simple addition, point by point, to re-create absolute coordinates that might be required by an output device such as a plotter or display.

XIII. ECONOMICS

The economics of the whole process must be considered as well as that of the various parts. The lowest overall costs can only be obtained by good planning and work control. In the following it has been an essential aim to reduce the amount of human labour to a minimum. The values below cannot at this time be proven by experience but a good argument can be made. It should be noted that a 'map' consists of multiple separations.

Human Involved Operation

	Description	Hours/ map	Equipment Cost/map	Equipment	Labour & Equipment Cost/map
1	Input inspection	1	$ 500	Light Table	$ 20
2	Output inspection	2	$100,000	Interactive Display and Edit	$ 80
3	Opaquing	1	$ 500	Light Table	$ 20
4	Contact printing	1	$ 10,000	Contact printer and processor	$ 22
5	Labelling	4	$100,000	Display and edit	$160
6	Correction	1	$ 20,000	Digitizing Table	$ 44
					$346

Automatic Operations

	Description	Hours/ map	Equipment Cost/map	Equipment	Labour & Equipment Cost/map
1	Scanning	1	$200,000	Scanner and processor	$ 40
2	Line Thinning	1/2	$100,000	Processor	$ 10
3	Line Following	1/2	$100,000	Processor	$ 10
4	De-symbolization	}			
5	De-zoning	} 1/2	$100,000	Processor	$ 10
6	Automatic Labelling	}			$ 70
					$486

Amortization, maintenance and labour are handled as follows:

1. 2,000 hours per year operation, 5 years life, i.e., 10,000 hours total
2. Maintenance costs equal to amortization
3. This results in $20/hour per $100,000 equipment cost
4. Labour costs at $20/hour

SPATIAL DATA INTEGRATION

Ellen M. Knapp

Geosciences Systems Department
Computer Sciences Corporation
Silver Spring, Maryland

Spatial data integration is defined as the process of combining multiple spatial data types and providing for their storage, retrieval, analysis, and display. Automated procedures to support data integration are related to spatial data characteristics, software features, and hardware capabilities. Spatial data integration issues associated with data characteristics are discussed as discipline issues; software and hardware features are referred to as automated data processing issues. Spatial data integration concepts, issues and research are reviewed.

I. INTRODUCTION

Increasingly, data producers are providing their spatial data in both analog and digital format. In addition, data users in government, academia, and private industry are converting existing data to digital files. These data are used to support a broad range of operational programs as well as research and development projects.

Spatial data are defined as recorded phenomena with implicit or explicit locational identifiers. This includes both spatial data content and spatial data entities. Content data are defined as recorded observations that are spatially distributed, which means that their values vary from one location to another, and data which relate to a single spatial entity, where variation in value occurs through time. The data may have been acquired directly by in-situ measurements or

ISBN 0-12-267180-5

indirectly by remote sensing methods. Spatial data entities are points, lines, areas, or surfaces with which content data are associated (e.g., weather stations, course lines, watersheds, and elevation models). Content data are assigned to spatial entities by locational identifiers. Three forms of locational identifiers are generally recognized: 1) external descriptive identifiers, 2) identifiers which assign content data to location within an array of discrete units, such as picture elements (pixels) or grid cells, and 3) coordinate identifiers which describe spatial entities as points, lines, areas, and surfaces in a continuous space, such as x, y coordinate strings or vector chains (1-3). Thus spatial data include content data and spatial entities, and may be four-dimensional: location in three dimensions (x, y, z) and time (t). The process of combining multiple spatial data types and providing for their storage, retrieval, analysis, and display is defined as spatial data integration (SDI).

Spatial data integration requirements were well established, before the implementation of automated procedures, by data users performing multivariate statistical or spatial analysis operations. As data handling requirements have progressed from data collection, mapping, and monitoring to predictive modeling and forecasting, the need for SDI capabilities has rapidly increased. For data users with large volumes of multiple spatial data types, automation can make a significant contribution to data processing in terms of cost and time required for data management, retrieval, analysis, and display in addition to providing capabilities which are not feasible, or not possible, using manual techniques. Unfortunately, however, many of the problems associated with manual data integration (i.e., data set compatibility, accuracy, reliability, and interpretive data analysis) in conjunction with complex problems specific to digital data, software, and hardware make automated data integration less than a straightforward set of procedures.

II. SPATIAL DATA INTEGRATION

Spatial data integration incorporates a wide range of data manipulation and data processing techniques which are directly related to data characteristics and software/hardware features. The relationship between spatial data characteristics (such as structure, format, and volume) and software/hardware features (such as storage, data manipulation algorithms, data management procedures, and peripheral devices) is established through data capture. Data capture refers to the collection of observable phenomena directly in machine-readable form and the transfer of non-digital source material to a machine-readable form which is usable in computer-aided data handling. Source material may include digital imagery, photographic products, maps, measurements, charts, tables, field notes, or published texts. It is important to recognize that all forms of data capture of directly measured or remotely sensed data involve both error and generalization. With source material which has been manually prepared (e.g., maps, charts, and tables), the original itself may include errors due to instrument characteristics or data preparation and compilation procedures. Although the focus of this review is on automated data handling, characteristics of manually prepared data sets which affect the accuracy and utility of SDI procedures will be considered (e.g., resolution, scale, accuracy, precision, and data organization).

Appropriate use of SDI procedures requires knowledge of the data characteristics themselves as well as their relationship to software/hardware features. Those integration issues which are associated with the data characteristics are referred to as discipline issues; those which are associated with software/hardware capabilities are discussed as automated data processing (ADP) issues. This is an essential, although frequently misunderstood, distinction. Data independent software and hardware features allow for the integration of spatial data in both appropriate and inappropriate ways. For example, using the same techniques, it is possible to display total population counts by census tract or population density by census tract. Many users are not aware of the major differences in the interpretability of these two products. The technique which allows for the retrieval and display of the data is an ADP capability; which measurement scale is appropriate for the display is a discipline issue. Unquestionably, the discipline issues are the less well defined. Most are well known to data producers (remote sensing specialists, cartographers, and statisticians) and are unknown to many data users; others have been identified by researchers and are, as yet, little understood. Nevertheless, the need for and the use of SDI techniques will not wait for a full set of ADP capabilities and completely defined discipline issues. Data users continue to apply existing methods, develop and implement new automated techniques, and should be as well informed as possible of the discipline issues related to the analyses performed. The following paragraphs briefly describe spatial data characteristics and software/hardware features as they relate to both SDI discipline and ADP issues.

A. Spatial Data Characteristics

As previously defined, spatial data include both content data and locational identifiers of spatial entities, and may be four dimensional. Characteristics of spatial data which affect the accuracy and utility of SDI techniques include the following:

- Medium on which the data are recorded
- Data organization (i.e., spatial data structure and data storage format)
- Scale
- Projection
- Coordinate system

- Observational criteria
- Measurement scale
- Remotely sensed data characteristics (i.e., sensor and platform characteristics, atmospheric effects, and resolution)
- Temporal characteristics (i.e., date, currency, and uniformity)
- Data volume
- Reliability (i.e., quality, accuracy, and precision)

General descriptions of these characteristics can be found in existing publications (4-11). A detailed analysis of the SDI discipline and ADP issues associated with the data characteristics is available in Spatial Data Integration I: Concepts, Requirements, and Current Capabilities (12). The following example of measuring temperature in the Earth's atmosphere illustrates some of these characteristics and their implications for spatial data integration.

Temperature may be measured directly by in-situ methods (such as direct measurement by a thermistor carried aloft by a weather balloon) or indirectly by remote sensing methods (such as inferring temperature from radiances detected by an infrared sensor onboard a weather satellite). In addition, the locational identifiers associated with the measurement may be two-dimensional, three-dimensional, or four-dimensional. Determining temperature at a specific place on the Earth's surface requires a three-dimensional identifier (e.g., latitude, longitude, and time). If the data required by the user are the minimum or maximum temperature during the month, then the temporal information may be suppressed and a two-dimensional identifier is sufficient. On the other hand, if the temperature profile (temperature as a function of altitude) is the desired spatial data, an additional spatial dimension is required to locate the data in altitude. The data collection method selected, the resultant data set characteristics, and the relative importance of each characteristic (i.e., accuracy, data organization, temporal characteristics) is largely a function of available data collection methods, the intended use of the data, and existing software/hardware capabilities. Further, these characteristics determine the applicability of SDI techniques. If the objective of temperature data collection is to compare and combine direct with indirect measurements, the inherent incompatibilities of each data type become limiting characteristics. A weather balloon (rawisonde) performs essentially a point measurement (fixed latitude and longitude) while a weather satellite observes radiation from a large area (typically 10 km on each side). It is well known that coincident profiles obtained by balloon and satellite do not agree. At least part of the discrepancy is due to fundamental differences in the data types measured and the paucity of point measurements available for comparison. Additional problems include data set registration (spatial and temporal), resolution and scale compatibility, data structure and format compatibility, and accuracy characteristics of the respective data sets.

Although a complete description of discipline issues is beyond the scope of this article, some significant SDI concerns related to spatial data characteristics are as follows:

1. Overlaying different data sets correctly requires knowledge of the fundamental differences in the data types measured and the incompatibilities in data characteristics (e.g., resolution, scale, data structure, and storage format).

2. A major SDI concern is comparability and compatibility of the organization of the data (data structure and data storage format). Conversion from one data structure to another may be so difficult as to be virtually impossible or may require information not present in the original structure.

3. Conversion between different coordinate systems may result in loss of information or the introduction of error as a result of interpolation, resampling, rescaling, or the round-off methods of individual machines.

4. Software to support quantitative assessment of the impact of scale changes, projection and coordinate system transformations, structure and format conversions, and data set reclassification or aggregation is not generally available within existing data handling systems.

5. Resampling, interpolation, and smoothing techniques are required for multi-dimensional data sets. Accuracy statements or confidence limits should be provided with output products. Interpolation and smoothing techniques for four-dimensional data sets have yet to be developed.

6. The type of observation (individual, grouped, or interpreted) and measurement scale affect the validity of information extraction techniques, statistical analysis procedures, data integration and mapping operations.

7. The approriate use of display devices requires an understanding of map design principles, interactive graphics, and the relationship of the symbols and colors to measurement scale so that no more information is conveyed than actually exists in the data.

8. To use indirect data effectively, the limiting features and characteristics of the sensor should be known. The atmosphere between the sensor and the object or phenomenon being sensed affects the measurement and should be taken into account.

9. Accurate registration of data sets is the most difficult and most frequently encountered SDI problem. Registration requires matching both geographic (locational identifiers) and temporal references as well as data characteristics (e.g., scale, resolution, projection, coordinate system, measurement scale).

Registration accuracy requirements are dependent upon the needs of the data user. For example, temporal registration may be necessary to within one second or to within a few years, depending upon the intended use of the data. Degradation of high resolution data and the concomitant loss of information is frequently the price of registration with low resolution data sets.

10. Overlay or comparison of remotely sensed data sets includes both multiple data sets from a single sensor (i.e., scene-to-scene registration) and data from several sensors. SDI issues related to scene-to-scene registration include spatial precision, temporal resolution, and the impact of atmospheric effects and sensor characteristics on the measurements for the different dates. These issues, in addition to different viewing and scanning geometry, different sensor characteristics, and different data characteristics, apply to registration of data sets from multiple sensors.

B. Software and Hardware Features

Automated spatial data integration has been defined as a set of procedures which are related to data characteristics and software/hardware capabilities. Examples of these capabilities include data manipulation algorithms, application packages, man-machine interfaces, data management systems, CPU and I/O requirements, storage type and limitations, and available peripheral devices. For convenience, software and hardware capabilities are discussed separately although they are interrelated. That is, an algorithm to overlay different types of spatial data must be tailored somewhat to its host machine. It must be small enough to fit into the available computer memory, be fast enough to fulfill its mission at a resonable cost, and not require peripheral devices other than those currently installed.

1. Software Issues. In order to be cost-effective for data integration, software development should incorporate certain basic precepts. For example, structured programming techniques and "top-down" design procedures can speed software development and subsequent enhancement. Similarly, software should be as transportable as possible. Although all software is to some degree machine dependent, transportability can be increased by the use of established software programming standards (regardless of the language selected), the use of standard input/output (I/O) procedures, and in-code documentation. Numerous problems arise from the transfer of spatial data handling programs from one computer to another. A program designed for efficiency on one machine may not be so on another due to different storage systems of arrays or transfer to a machine with a virtual operating system. Programs transferred to a machine that utilizes parallel processing may require complete restructuring.

Additional software related issues are data management, data structures and formats, human engineering, and documentation. The large volumes of data, variety of data types (with associated structures and formats), and storage media which must be accomodated to support SDI combine to make data management the largest software problem. Efficient data storage, retrieval, and update are, to a large extent, dependent upon the data organization, storage medium, and hardware features such as memory size and I/O procedures. Major efforts are being directed toward the development of spatial data base management systems (13-15).

The proliferation of spatial data structures and formats poses significant software problems (12). Routines are necessary to convert data from one structure to another in order to accomodate various software packages and peripheral devices. This is an on-going, time-consuming, costly SDI requirement given the number of existing structures and formats, the differences between them and the fact that new forms of data organization are continually being developed and implemented (16-19).

Human engineering usually refers to the way in which the software user interacts with the software (inputs controls or data and receives program output). Most problems encountered using existing software are related to human failings (e.g., inputting the wrong control cards). Human engineering considerations can reduce these, but current attempts are still somewhat primitive. For example, the use of a standard, readable menu-type man-machine interface is a major requirement which has not yet been achieved. Finally, documentation -- adequate, readable documentation -- continues to be a major software related problem.

2. Hardware Issues. Hardware features affect software design and transportability, processing efficiency, I/O procedures, and display capabilities. These features present complex problems for data users because of the variety of different computers and peripherals currently available to support automated data handling. The computers often have different word sizes, machine instruction sets, software utilities, and processing modes (batch or interactive). Minicomputers require programmers to modularize, overlay, and multi-task, but these processes often involve designs which are unique to a single system, are hardware dependent, or rely on specific devices. Transporting a spatial data manipulation program from a mainframe to a minicomputer is often a difficult task even without taking into account special hardware items existing on many dedicated minicomputers; transporting a program from a minicomputer to a mainframe frequently involves the removal of multi-tasking.

The use of specialized hardware provides additional data handling capabilities but may introduce additional problems. Data capture devices, ranging from hand-held digitizers to laser scanners, each produce their own unique data structure and format. Array processors and other specialized hardware for parallel processing affect data organization, software design, and processing efficiency. Furthermore, special output devices such as storage tube and raster CRTs, vector and raster plotters, and film recorders require device or system dependent formats and software utilities.

Although hardware characteristics are primarily ADP related, discipline issues such as data storage and processing accuracy may be affected as well. For example, data compaction techniques employed to accomodate limited storage space or machines with small word sizes may introduce error in data sets (20). Similarly, rounding-off procedures may affect data manipulation and analysis results (e.g., coordinate transformations, area measurements, or statistical analyses). Hence thc efficient use of existing capabilities requires expertise in many unrelated fields. Familiarity with software and hardware characteristics is an important requirement too frequently overlooked or ignored by spatial data users.

III. SUMMARY

Automated spatial data integration is a complex process which requires an understanding of spatial data, data characteristics, data collection and data capture methods, software characteristics, and hardware features. ADP issues such as data storage, data management, processing efficiency, and format conversion present a constant challenge in spatial data handling. Considerable efforts are currently directed toward addressing ADP issues (i.e., improving the utility of existing software and hardware and developing additional capabilities). As spatial data users learn to communicate their objectives more effectively, computer scientists, engineers, and programmers respond with innovative concepts in software and hardware design (e.g., fast algorithms, relational structures for spatial data base management, raster scanning devices, and future sensor designs). Unquestionably, this research and development is needed to improve problem solving and decision-making capabilities of data users. Simply addressing ADP problems, however, is not sufficient. The quality and reliability of processing results -- and thus the utility of output products -- will be determined by the way in which discipline issues are treated. Many issues related to data characteristics such as measurement scale have been identified by researchers and should be applied in data analysis. Additional issues have not yet been sufficiently addressed; for example, decision rules for assessing data set compatibility and improved accuracy evaluation procedures are required. As environmental analysis and modeling activities expand, requirements for automated data handling techniques, including SDI, will rapidly increase. Greater efforts should be directed toward understanding and communicating the impact of both ADP and discipline issues through research, publication, and standards development.

ACKNOWLEDGMENTS

The author would like to thank those who have assisted in the preparation of this article and supported spatial data integration studies, especially Scott Turner and Ron Nieman of Computer Sciences Corporation and Richard desJardin and Walter Truszkowski of the National Aeronautics and Space Administration. Special thanks go to Professors Freeman and Pieroni for their organization and direction of the NATO Advanced Study Institute.

REFERENCES

1. Chrisman, Nicholas R., "Concepts of Space As a Guide to Cartographic Data Structures," Papers of the Advanced Study Symposium on Topological Data Structures for Geographic Information Systems, Vol. 2, Harvard University, 1977.

2. Peucker, Thomas, K., Computer Cartography, American Association of Geographers Commision on College Geography, Resource Paper No. 17, Washington, D.C., 1972.

3. Freeman, H., "Application of Generalized Chain Coding Scheme to Map Data Processing," paper presented at the IEEE Computer Society Conference on Pattern Recognition and Image Processing, Chicago, 1978.

4. Calkins, H.W., and Tomlinson, R.F., Geographic Information Systems, Methods, and Equipment for Land Use Planning, International Geographical Union Commission on Data Sensing and Processing, Ottawa, 1977.

5. Robinson, A., Sale, R., and Morrison, J., Elements of Cartography, New York: John Wiley & Sons, Inc. Fourth Edition, 1978.

6. Peucker, Tom K., and Chrisman, Nicholas, "Cartographic Data Structures," The American Cartographer, Vol. 2, 1975.

7. Richardus, Peter, and Adler, Ron K., Map Projections, Amsterdam: North Holland Publishing Company, 1974.

8. Snyder, John P., "Map Projections for Satellite Applications," Proceedings of the American Congress on Surveying and Mapping, 39th Annual Meeting, 1979, pp. 134-145.

9. Davis, Peter, Data Description and Presentation, Science in Geography (3), London, Oxford University Press, 1974.

10. Lintz, J. Jr., and Simonett, D., editors, Remote Sensing of the Environment, Reading, Massachusetts: Addison-Wesley, 1976.

11. Reeves, R. G., Anson, A., and Landen, D., editors, Manual of Remote Sensing, Falls Church, Virginia: American Society of Photogrammetry, 1975.

12. Knapp, E., et al., Spatial Data Integration I: Concepts, Requirements, and Current Capabilities, CSC/TM-79/6241, Computer Sciences Corporation, Silver Spring, Maryland, 1979.

13. Tomlinson, R. F., "Difficulties Inherent in Organizing Earth Data in a Storage for Query," Proceedings of the Third International Symposium on Computer-Assisted Cartography, San Francisco, 1978, pp. 181-201.

14. Business and Technological Systems, Inc., A Conceptual Design for an Integrated Data Base Management System for Remote Sensing Data, BTS-FR-78-65, Maresca, Paul A., and Lefler, R. Michael, Seabrook, Maryland, 1978.

15. Shapiro, Linda G., "Design of a Spatial Information System," presentation at the NATO Advanced Study Institute on Map Data Processing, Maratea, Italy, 1979.

16. Cederberg, Roger L. T., "Chain-Link Coding and Segmentation for Raster Scan Devices," Computer Graphics and Image Processing, 1979, pp. 0128-0138.

17. Boyle, A. R., "Scan Digitization of Cartographic Data," paper prepared for the NATO Advanced Study Institute on Map Data Processing, Maratea, Italy, 1979.

18. Nagy, George, and Wagle, Sharad G., "Approximation of Polygonal Maps by Cellular Maps," Communications of the ACM, Fritsch, F.N., editor, 22:9, 1979, pp. 518-525.

19. Rosenfeld, Azriel, "Extraction of Topological Information from Digital Images," Papers of the Advanced Study Symposium on Topological Data Structures, Harvard University, 1977.

20. Rogers, David F., and Adams, J. Alan, Mathematical Elements for Computer Graphics, New York: McGraw-Hill, Inc., 1976, p. 5.

A SPATIAL DATA STRUCTURE FOR GEOGRAPHIC INFORMATION SYSTEMS

Robert M. Haralick

Department of Electrical Engineering
Virginia Polytechnic Institute and State University
Blacksburg, Virginia

ABSTRACT

In this chapter we describe a data structure that is equally natural for spatial data in the raster run length format and the vector format. This makes it easier to design a geographic information system that handles both data formats and that optimizes for itself whether any particular data set is in raster or vector format. This same data format is appropriate for intelligent query and inference uses as described in Shapiro's chapter of this book.

Brief mention is made of the topological consistency requirements for each data format. A bibliography of many of the fast algorithms for vector format data is provided at the end of the paper.

I. INTRODUCTION

Geographic information systems have requirements for high volume spatial data handling. Because the volume of spatial data is so high, it is particularly important for the geographic information systems to have efficient data storage formats and associated optimal algorithms.

ISBN 0-12-267180-5

Spatial data has two natural organizations: raster format and vector format. Arguments have been made in favor of each of them. It is the thesis of this paper that the particular format and associated algorithms by which a user's spatial data is processed is not relevant to the user. What is relevant is the cost of processing. This suggests that the geographic information system must be able to handle both the raster and vector formats using the particular format most cost effective for each individual's spatial data processing profile and spatial data set. Format conversions, if determined to be cost effective for a processing profile, are the responsibility of the geographic information system, not the user.

Because the vector data format and the raster data format are so different, one might expect that there is no logical data structure that naturally can contain these formats. From this it might follow that a geographic information system which had to use and keep track of the different formats would be excessively complex. It is the purpose of this paper to argue against such a conclusion.

We first discuss a logical spatial data structure which naturally and efficiently can contain the vector format data or raster format data. Second we describe some of the efficient processing algorithms associated with each of the formats. Note that the structure we discuss here is one that is suitable for the inference algorithms that enable high level queries to be answered. Its use in inference can be found in Shapiro's chapter of this book.

Rather than begin with a definition of the suggested spatial data structure, we begin with a toy example which has in it perhaps just enough complexity to indicate the universality of the data structure for spatial data processing as well as inference work. The example emphasizes spatial relationships without exclusively using coordinate information. This kind of information is sometimes useful in performing limited inference tasks which do not have to depend on coordinates. In the example, we will give a written description of some spatial information about a place called Ecaf. Then we will encode it in a brief tabular form closely related to our logical spatial data structure and illustrate a map of Ecaf visually portraying the spatial information. Finally we will put the information about Ecaf into our suggested spatial data structure.

Example

Ecaf has a population of 1,252,140 and occupies a near circular area of 30,179 square mm. It is located in the uppermost part of Ydob. It has four main regional centers of activity: htuom, eson, and the twin seyes.

The twin seyes are located in the northern part of Ecaf, one in the west side and one in the east side. Each was planned and constructed identically by the same master architect. Each occupies a 200 square mm oval shaped area, elongated in the east west direction. Each has a circular downtown business area of about 20 square mm located in its west end. Lying on low land, east and west seye are protected on the north from the seasonal flooding by a barrier forest. The main function of the seyes is to act as the receiving center for the high bandwidth information that comes into Ecaf.

Eson is a city of 150,000 located in the center of Ecaf.

It has a thin triangular 300 square mm area elongated in the north-south direction and its northern tip is between the two seyes. It functions as the gas trade center of Ecaf.

Htuom is a city of 200,100 located in the southern part of Ecaf. It has a thin elongated 1500 square mm area stretching from southwest Ecaf to southeast Ecaf. Its east and west ends have been experiencing northward growth. Htuom is an industrial city importing raw materials as well as being an outgoing communication center of Ecaf.

Tables 1 through 6 list the information about Ecaf in tabular form. Some of the data such as population or area are just characteristics of Ecaf. Other data such as boundary give the name of the set of coordinates defining the boundary of the areas involved. Figure 1 gives a map of Ecaf, a familiar place: the human face.

We will now classify the kind of data in the tables. First, every table contains the name and type of the entity represented by that table. Entity type provides the key to the entity type prototype which indicates the kind of information that can be in the entity type and the meaning of the information. Shapiro's chapter in this book fully describes the notion of prototype. Second the table contains the attributes the entity can have; shape, area, population, and so on. Attributes can have values which are atomic (numbers, character strings, or N-tuples) or names of other entities. Finally,there are relations which may be simple lists (ordered or unordered) such as the unordered list (a unary relation) associated with "contains". A relation may be a set of ordered pairs such as the binary relation of coordinates associated with "boundary", or may be more complicated N-ary relations.

Entity Name	Ecaf
Entity Type	County
Population	1,252,140
Area	30,179 sq mm
Area Shape	Near Circular
Situated In	Northern Part of Ydob
Contains	Htuom, Eson, and Twin Seyes
Boundary	Ecaf Boundary

Table 1 lists the information available about Ecaf

Entity Name	Twin Seyes
Entity Type	City
Situated In	Northern Ecaf
Contains	Seyes East and Seyes West

Table 2 lists the information about Twin Seyes

Entity Name	Seyes East
Entity Type	City
Population	279,300
Area	200 sq mm
Area Shape	Oval, Elongated in East-West direction
Situated In	Northeast Ecaf
Contains	Downtown Seyes East
Boundary	Seyes East Boundary
Protected By	Forest East
Located On	Lowland
Functions As	Receiving Center

Table 3 lists the information about Seyes East

Entity Name	Downtown Seyes East
Entity Type	Central Business District
Area	20 sq mm
Area Shape	Circular
Situated In	West Side of Seyes East

Table 4 lists the information on Downtown Seyes East

Entity Name	Eson
Entity Type	City
Area	300 sq mm
Area Shape	Thin Triangular, North-South Elongated
Situated In	Central Ecaf
Contains	Twin Slartsons
Boundary	Eson Boundary
Functions As	Gas Trade Center

Table 5 lists information on Eson

Entity Name	Htuom
Entity Type	City
Population	200,120
Area	1500 sq mm
Area Shape	Elongated
Situated In	Southern Ecaf
Functions As	Industrial Center
Boundary	Htuom Boundary

Table 6 lists the information on Htuom

Figure 1 shows a map of the human face.

This suggests the following formal definition of a general spatial data structure: A spatial data structure S is a set of named relations $\{R_1,\ldots,R_k\}$ where each relation R_k is some N_k-ary relation whose tuples may have components whose values are atoms or spatial data structures.

To illustrate this structure, Figure 2 shows the Ecaf information in Table 1 put into the logical form of the spatial data structure. The entity name and type information actually appear in the directory of the spatial information system. The purpose of the directory is to give the "address" of any spatial data structure as well as the name of its prototype. Ecaf consists of four relations: an attribute value table, a contains relation, a situated relation, and a boundary relation. The attribute value table is a binary relation and has the attributes population, area, and shape with their corresponding values. The contains relation is a unary relation consisting of the simple set (unordered list) Htuom, Eson, and Twin Seyes. The situated in relation is a binary relation consisting of the set whose only member is the pair (northern, ydob). The character string "northern" means north part of and is really the name of a generic relation. The boundary for Ecaf is assumed to be a polygon specified by its vertices an ordered set of (x,y) coordinate pairs. We may think of this as an ordered binary relation or as a ternary relation.

Figure 2 only shows the logical structure. There are a wide range of choices for the physical implementation of this logical structure. Reasonable physical implementations would probably not explicitly represent entity name and type, attribute value table, contain, situated in, and boundary as

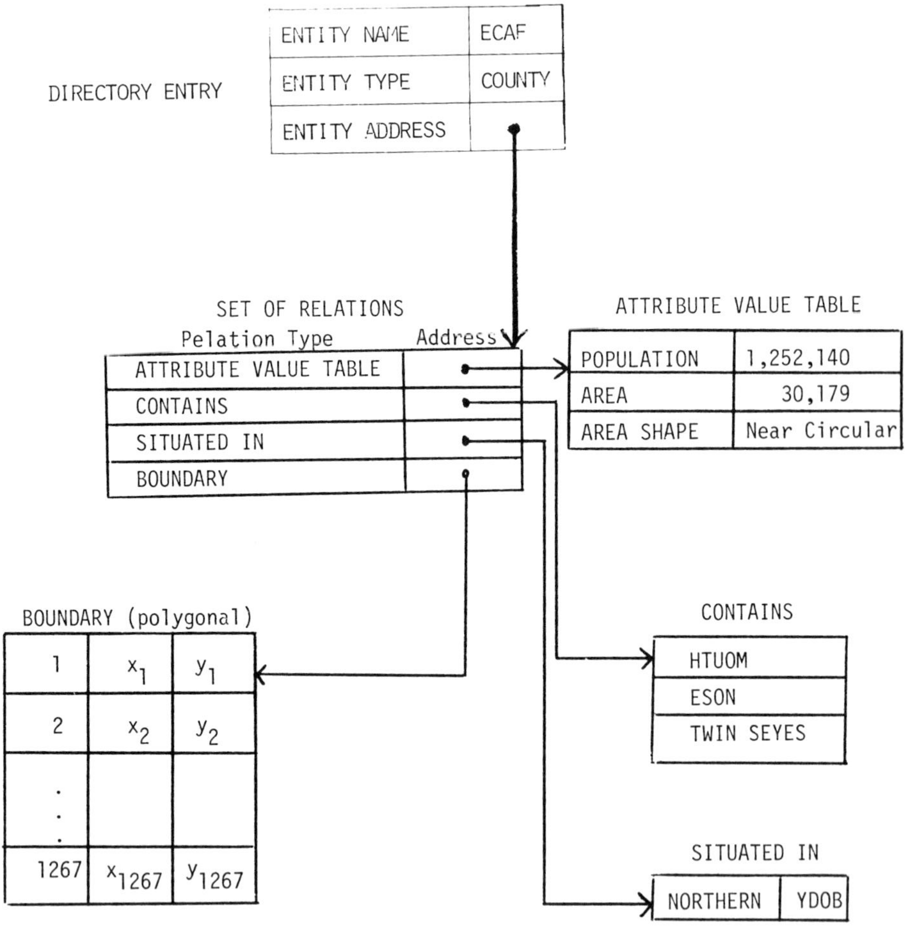

Figure 2 shows the Ecaf information of Table 1 put into the logical form of the spatial data structure character strings. Conventions could establish the order in which entity name, type, and address occur in a directory. The types of relations, associated with their addresses, could be given by bit table, for example. The ordered relations such as boundary would not explicitly represent the indexing giving the order of the (x,y) coordinate pairs. The order would be implicitly in the natural ordering of words in memory.

II. VECTOR FORMAT SPATIAL DATA

In vector format spatial data, areas are represented by their boundaries. Simple boundaries are simple closed curves which separate the inside of an area from its outside. To enable the distinction of inside and outside to take place, we must consider the closed curve to be directed. As we travel on the closed curve in the designated direction there will be an area to our right and an area to our left. Each of these areas can be associated with their names, thereby allowing the designation of inside and outside to occur. See Figure 3.

In complex spatial data, the total area is divided into many regions with many pairs of regions sharing some portion of their boundary. If we follow the data organization principle that each data item may only be represented once, then the closed curve bounding any area must be divided up into segments, each segment being a connected boundary arc whose left side is one region and whose right side is another region. In this manner each region can reference the same arc segment which is part of a shared boundary. Figure 4 illustrates a map composed of four regions, A, B, C, D, and with directed arcs a through m bounding these regions. Table

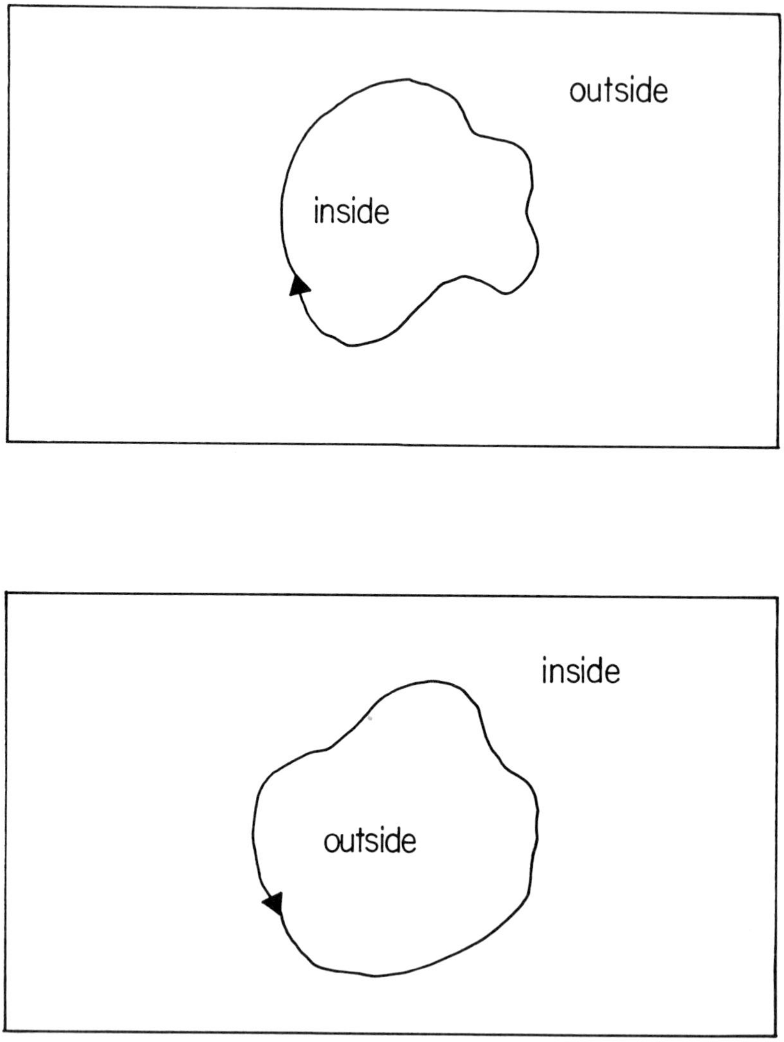

Figure 3 illustrates the directed closed curve which can label areas. In this example, the area to the right is the inside area.

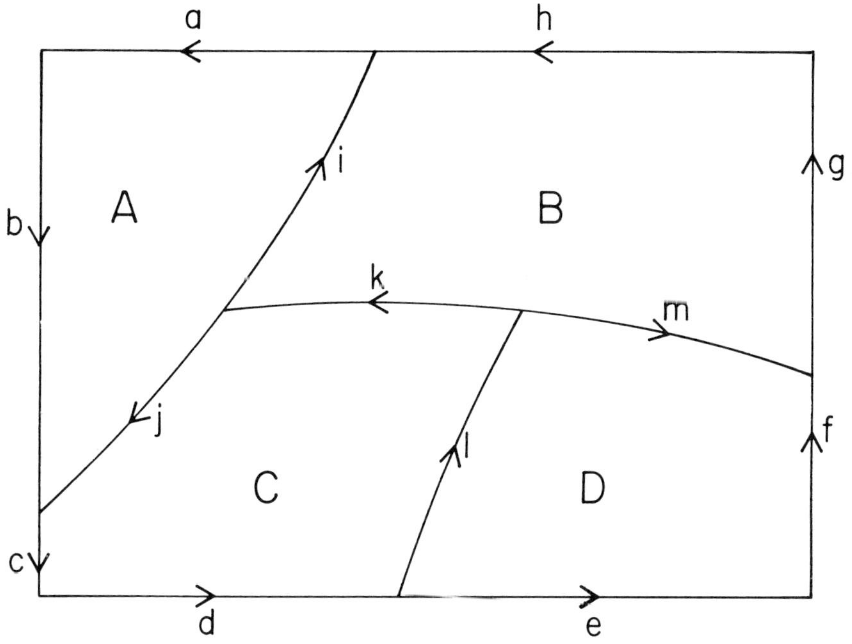

Figure 4 illustrates a map composed of four regions with directed arcs bounding these regions.

7 lists the relationship between arc name, its left region, and its right region for this example.

The boundary for region B then consists of the ordered sequence of arcs i, h, g, m and k. Traversing these arcs in the sequence given produces region B to the right of the traversed direction. Notice that at the place where two arcs meet it is not necessary that the arc directions be the same. Also, the direction in which we travel around a closed boundary does not have to agree with the direction specified in any of the arc segments of the boundary. The arc segment directions only give meaning to the region at the arc's left or right. From this example we see that a simple boundary is an ordered set of arcs forming a simple closed curve.

Another characteristic of complex spatial data is the fact that there may be some regions which have holes (regions which they entirely surround). Furthermore, the holes themselves may have holes which are part of the region. Such a region is necessarily disconnected and the determination of whether a point is in the region or not must make full use of the boundary information of the holes. Figure 5 illustrates a simple example.

This motivates the following tentative definition of boundary. A boundary is a simple boundary plus a (possibly empty) set of boundaries of the holes for the region it bounds.

This definition would be alright except for the fact that a region may have disconnected parts no one of which is a hole in another as illustrated in the map of Figure 6. This suggests that a boundary is a set of simple boundaries, each one of which bounds a connected region and has an associated

ARC NAME	REGION LEFT	REGION RIGHT
a	A	--
b	A	--
c	C	--
d	C	--
e	D	--
f	D	--
g	B	--
h	B	--
i	A	B
j	C	A
k	C	B
l	C	D
m	B	D

Table 7 shows the association of region left and region right to each directed arc of figure

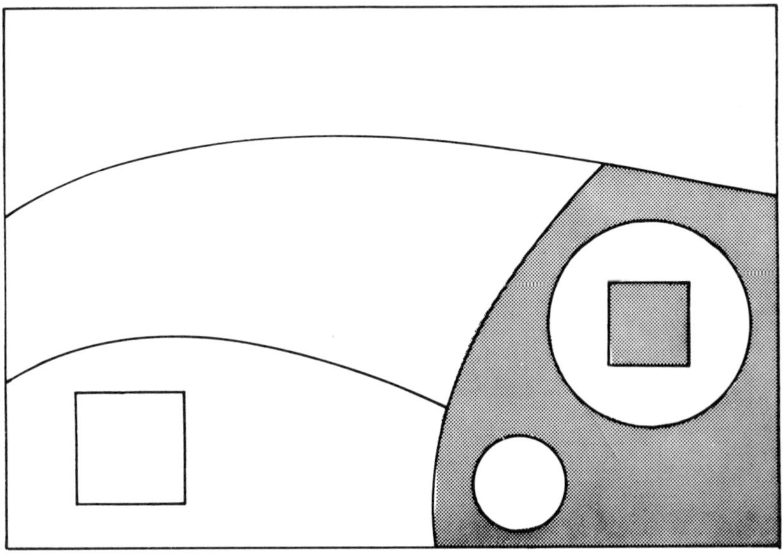

Figure 5 illustrates an example where regions have holes. Notice that one of the holes itself has a hole which is part of the region.

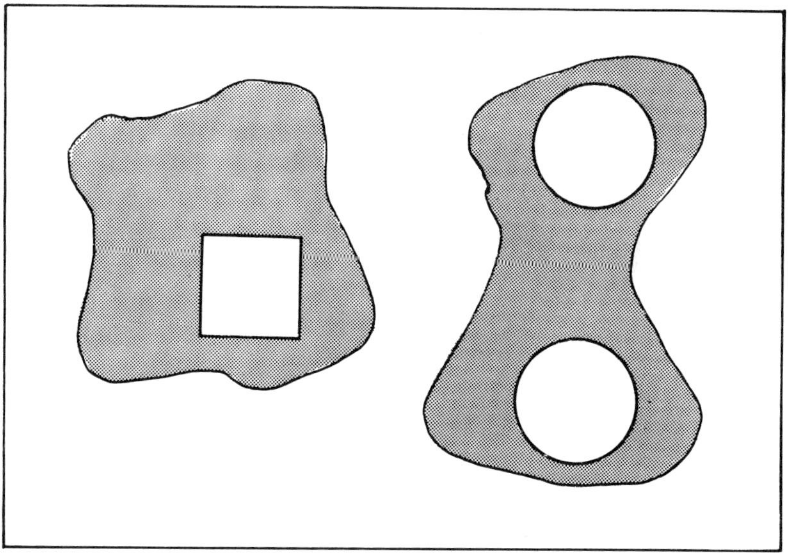

Figure 6 shows a map having a region which is disconnected and each connected part having at least one hole.

(possibly empty) set of boundaries for the holes of the connected region it bounds.

The definition of a boundary then is perhaps more complex than one might initially think. But it does bring out quite nicely a very important and interesting property of spatial information; spatial information at a high level is relational and at a lower level is more data coordinate oriented. Boundary is not just a simple ordered binary relation of (x,y) coordinate pairs. Boundary is a full and complete spatial data entity in its own right.

Illustrating this idea with an example from Ecaf, consider the boundary of Twin Seyes. Twin Seyes is a disconnected region with two pieces Seyes East and Seyes West. Seyes East and West are connected regions each having a simple boundary and each having a hole. Let us suppose that the name of the boundary for Twin Seyes is Twin Seyes Bndry. It is a spatial data structure having an attribute value table and Boundary of relation. The Boundary of relation is a unary relation, a set, whose members are Seyes East and Seyes West. These are, of course, spatial data structures of regions having boundaries. As can be seen from Figure 7 the spatial data structure for Seyes East has a unary relation of type B. The singleton member of this set is Seyes East B, the name of the spatial data structure for the boundary of Seyes East. Referring to Figure 8, Seyes East B is a spatial data structure having for its relations an attribute value table, a simple boundary, and a set for the boundaries of its holes. The simple boundary is an ordered set of five arcs.

The circularly ordered list of five arcs may actually be

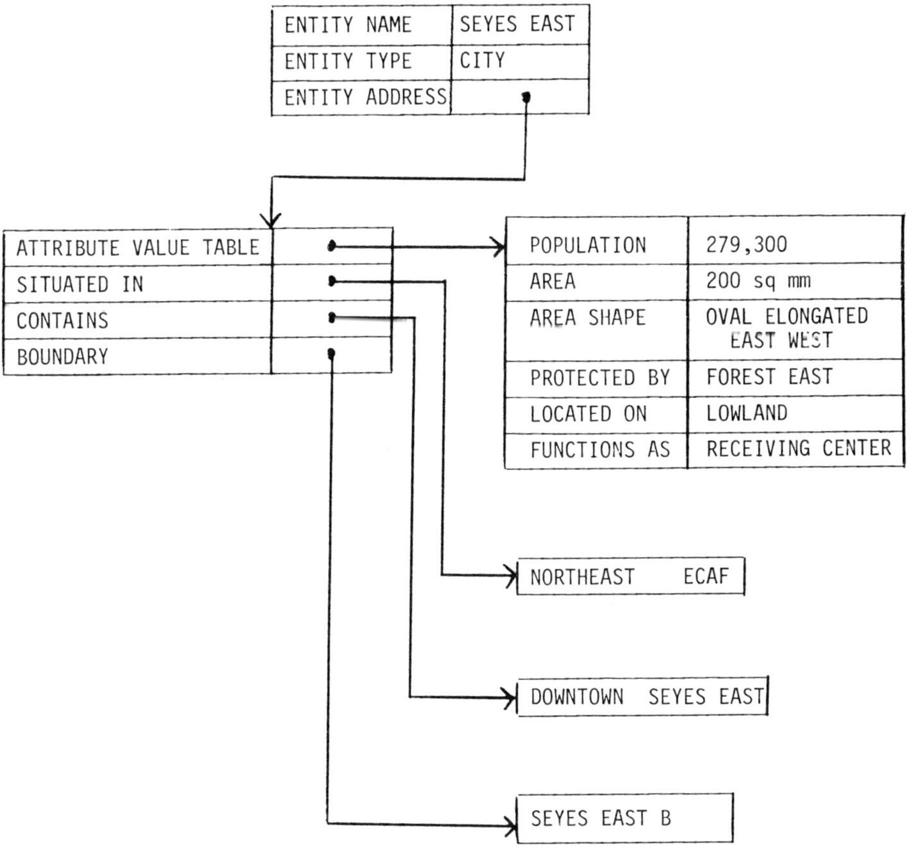

Figure 7 illustrates the logical form of the spatial data structure for Seyes East

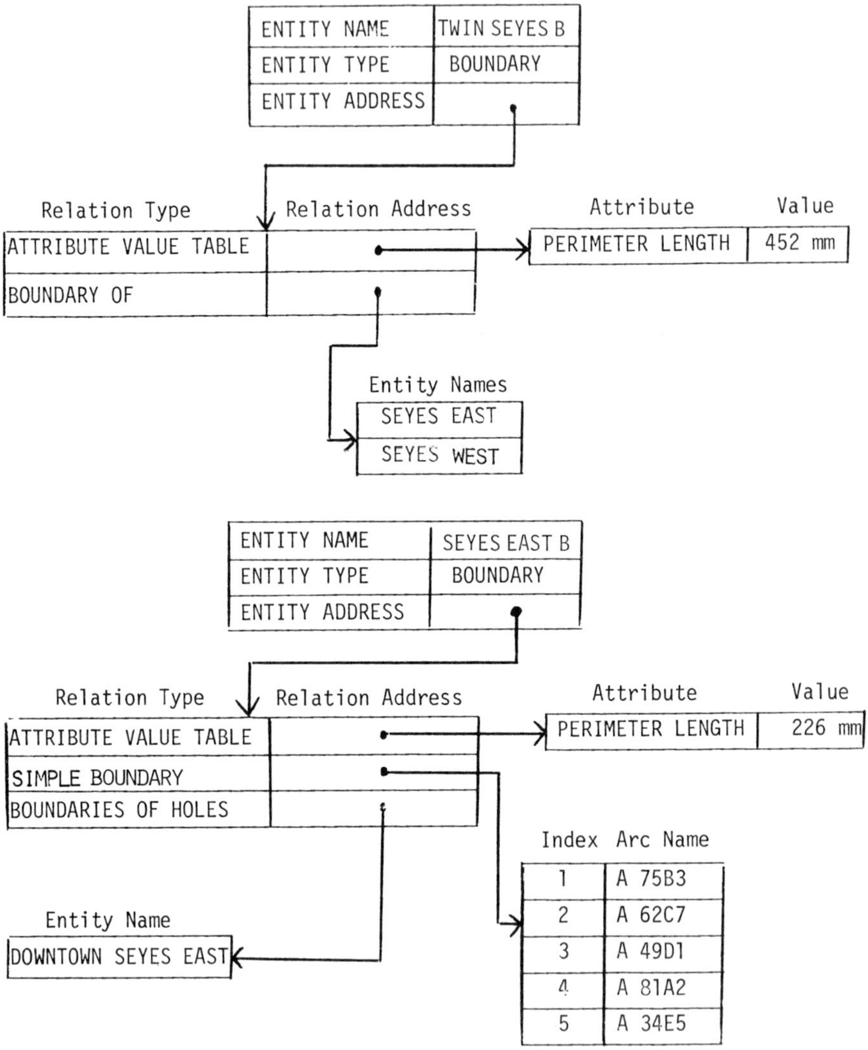

Figure 8 shows the logical structure for the boundary of the disconnected region called Twin Seyes in the Ecaf example.

stored in the logical form shown in Figure 8 or be stored by having each arc point to its next arc left and next arc right as in Tomlinson's CGIS system. In this latter case, the relation type "Simple Boundary" points to any arc in the sequence of 5 with a designation as to whether the region is to the arc's left or right.

II.1 Topologic Consistency

One of the important and non-trivial operations that must be performed on any vector format spatial data is topologic consistency. Topologic consistency demands that:

1) every point in the plane be in some region bounded by a sequence of arcs in the data base; and
2) there are no extraneous arcs in the data base.

There are two ways of performing consistency checks. In the first way, all entities must be searched for all boundaries. In each simple boundary, it must be verified that in its circularly ordered list of arc names, one end point of every arc must match one end point of the successor arc. As each arc in succession is being checked, reference can be made to the (arc, region left, region right) relation to verify that the current region being circled indeed appears to the arc's right when traversing the arc in the direction of the ordered arc sequence bounding the region. Also, each arc must have some region which references it and is to its left and must have some region which references it and is to its right.

The second way of topologic consistency checking requires that a list of arc end points be prepared. In this list, each end point, called a node, must point to all the arcs having it for one of its end points.

The consistency checking is accomplished by searching through all nodes. For each node begin with an arbitrary associated arc. Then circle the node by jumping from arc to arc. Walking along the arc in a direction towards the node, determine who the region on its left is. Then search the list of remaining arcs to determine which arc has that region on its right (again, taking the sense of direction to be that of walking toward the node). Repeat the process of jumping from arc to arc until the first arc selected is returned to. There is a topologic inconsistency if we cannot return back to the initial arc. And if all arcs have not been touched, then there is either garbage arcs or possible errors. Since each arc has two ends, each arc will be accessed twice in this scheme too.

III. RASTER FORMAT SPATIAL DATA

We assume that the raster data is compressed into a run-length variant known as Y partitions (Merrill, 1973). In this data structure, for each connected region, we must store the first row r_1 and the last row r_N having some pixel in the region. Then for each row r, $r_1 \leq r \leq r_N$, we must maintain an ordered list l(r). Each element of l(r) is a maximal horizontal strip which is contained in the given region. The list l(r) has as elements column pairs of the form (b,e), where b designates the beginning pixel and e designates the ending pixel of an interval in the region. Thus, if $(b,e) \in l(r)$, then $\{(r,b), (r,b+1), \ldots, (r,e)\}$ must be contained in the region, and (r,b-1) and (r,e+1) do not lie in the region.

Figure 9 illustrates Seyes East again, but this time using the raster format data representation. Notice that

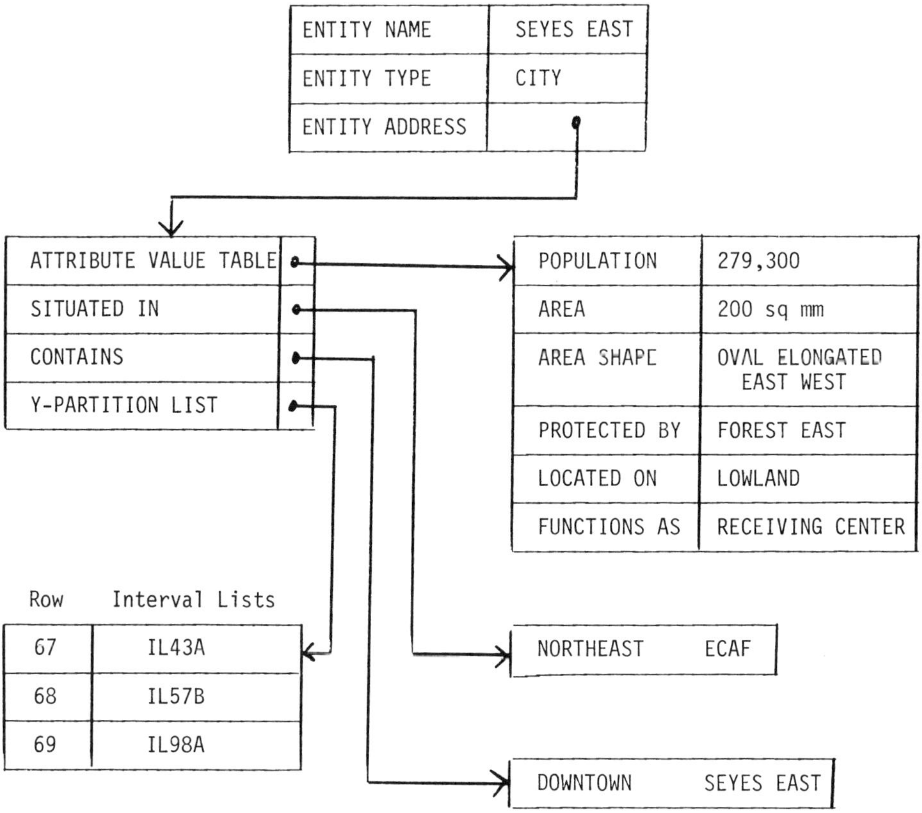

Figure 9 illustrates the logical form of the spatial data structure for Seyes East when the data format is raster except for the replacement of boundary by partition list, little has changed compared to Figure 7. Figure 10 illustrates the logical structure of a Y-partition list. It is shown as a binary relation having three (row index, interval lists) pairs. Each interval list is a spatial data structure itself containing its intervals as a binary relation of ordered pairs whose first component is beginning pixel and whose last component is ending pixel of an interval. In its attribute value table is its associated region name. Finding all the pixels in a given region is done by going through all the interval lists in the Y-partition of the region. Each interval list specifies pixels contained in the region.

Another convenient data structure for raster format data is a master set of intervals relation shown in Figure 11. This relation stores for each row all the interval lists in the row. The master set of intervals relation can be created from the Y-partitions in the following way. Search through all regions and their Y-partitions. For each Y-partition pick up the interval lists and their corresponding row indexes. Simply add the intervals in the interval list to the set of intervals in the master on the corresponding row.

Finding what region a pixel (r,c) belongs to is done by consulting the master list under row r and searching all the associated intervals to find that one containing c. The attribute value table then gives the region name.

Verifying topologic consistency amounts to verifying that for every row r of the master list, the associated set of intervals constitutes a partition of the pixels in row r. Pixels which are missing from all intervals in a row or which appear in more than one interval are errors.

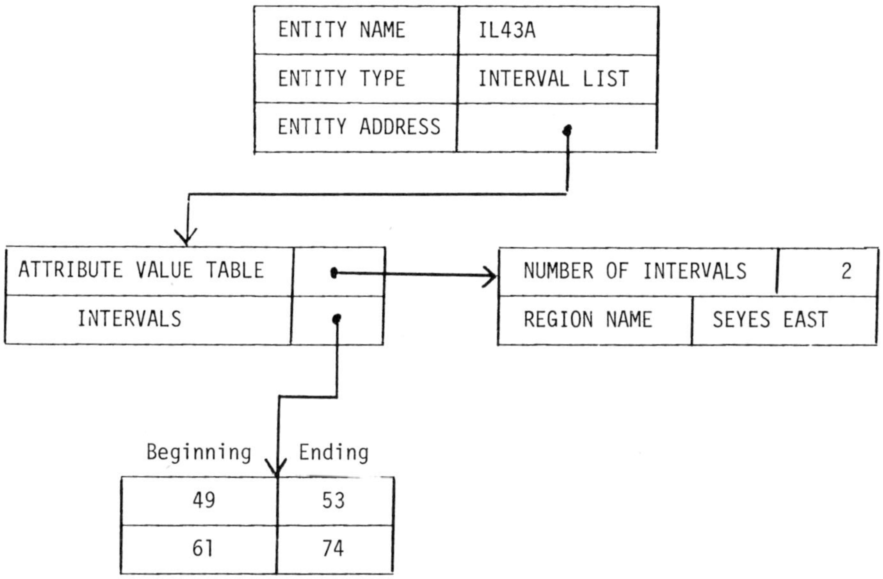

Figure 10 illustrates the logical structure of the interval lists

Row	Interval List
1	IL 10Z
1	IL 15A
1	IL 9B
2	IL 17C
. . .	. . .
67	IL 43A
. . .	. . .
512	IL 77D

Figure 11 illustrates the master interval list relation data structure

IV. RASTER FORMAT VS. VECTOR FORMAT

Raster format data has an advantage over vector format data in that the algorithms to determine in what region a pixel lies,to determine the boundary of a given region or to do polygon overlay, are simpler algorithms than the corresponding algorithms for vector format data. However, it is not necessarily the case that the simpler algorithm for raster format data will execute faster than the corresponding algorithm for vector format data because raster data typically has a much higher data volume. Which algorithm will execute quicker depends on the characteristics of the data set being operated on. The thesis of this chapter is that the data format question and its associated algorithm should not be the worry of a user of a spatial information system. Rather it should be the worry of the system itself.

This point of view which admits the possibility of data sets executing quicker in vector format is possible because of the recent advances made in the field of computational geometry. This research has shown that the typical brute force searching algorithms for vector format data are not the most efficient. More efficient algorithms exist. Each finds some way to use the order in a data point set to increase the search efficiency.

A bibliography of papers discussing and describing these fast geometric search algorithms for vector format data is given at the end of this chapter. Because a complete review of these algorithms would be quite lengthy, in the next few paragraphs we give the reader a feel for them by describing only one of them, an algorithm to solve the point in region problem published by Lee and Preparata (1977).

The point in region problem is given a point, determine which region it is in. The worst algorithm to use in solving this problem is the simplest one: search through all regions and determine the first region in which the given point lies. The algorithm described by Preparata and Lee is based on organizing the line segments of the region boundaries into monotone chains.

A chain of line segments is said to be monotone with respect to a line ℓ if the order of the verticies of the chain and the order of the vertices of the chain as projected onto line ℓ are identical. Figure 12 illustrates an example of a monotone chain and a non-monotone chain.

A quick algorithm exists for determining on which side of a monotone chain (with respect to the line ℓ) a point lies. Take the projection of the point to the line ℓ. The monotonicity of the chain implies that each interval on the projected line will correspond to exactly one line segment of the chain. Determine that line segment containing the projected point by a binary search and solve the problem on which side of the line segment does the point lie. The point lies on the same side of the chain as this side of the line segment.

Preparata and Lee decompose the set of boundary line segments into a monotone set of chains. In this set each chain is monotone with respect to the same line, each boundary line segment belongs to at least one chain, and the vertices of a chain C which are not in a chain C' all lie on the same side of C'. Figure 13 illustrates a set of regions decomposed

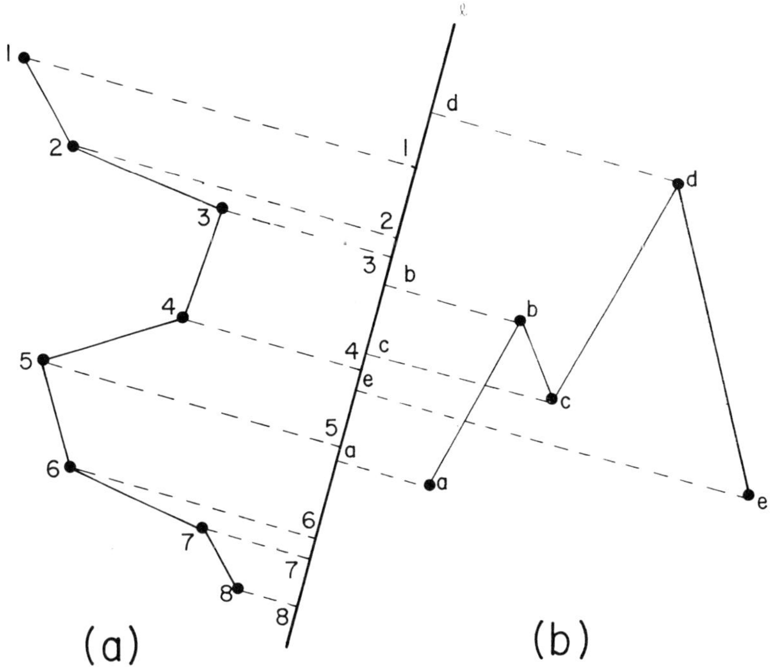

Monotone with respect to ℓ

Not monotone with respect to ℓ

Figure 12 illustrates a monotone and non-monotone chain with respect to the line ℓ

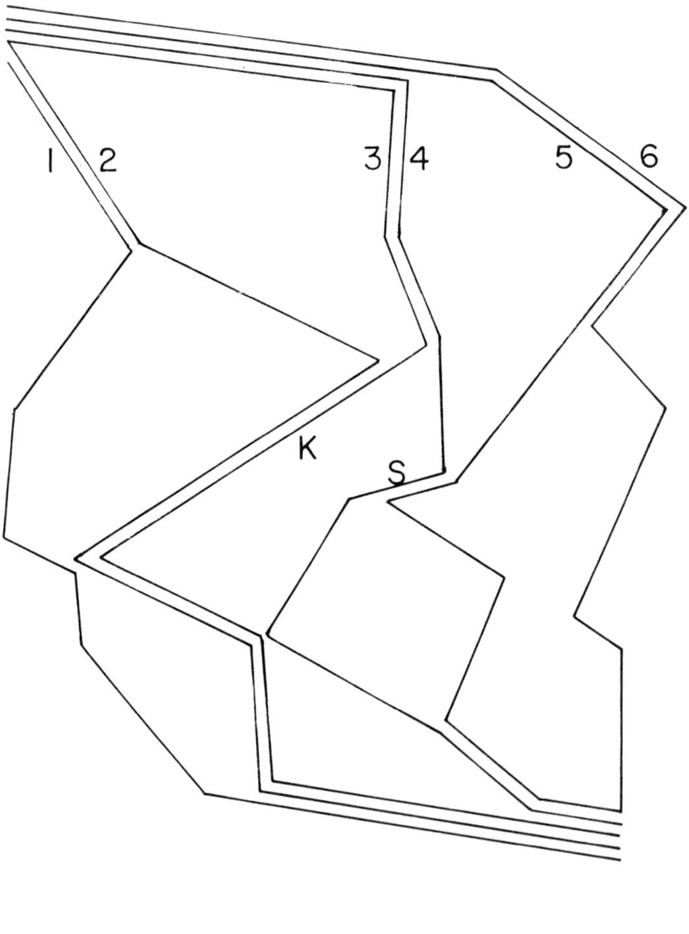

Figure 13 illustrates a set of regions decomposed into 6 chains monotone with respect to the y-axis.

into 6 chains all monotone with respect to the y-axis. Notice that because the chains are not allowed to cross one another (although they are allowed to share line segments), the chains can be simply ordered in a left to right fashion.

To determine in what region a point lies, do a binary search on the chain set. Begin with a chain in the middle and determine whether the point is to the left or right of the middlechain. Continue the binary search until it is determined that the point lies between two successive chains. Then take the left chain and obtain the line segment of the chain having a point with same y-coordinate as the given point. If the line segment is directed upward, the desired region is on the line segment's right. Otherwise it is on its left.

Figure 14 illustrates the binary nature of the search tree. If there are K chains, the binary search will search through $\log_2 K$ chains to find the two chains the given point is between. If there are N vertices per chain, each side of chain search will take $\log_2 N$ operations. Thus, the number of operations required to do the point in region problem is $(\log_2 K)(\log_2 N)$.

Preparata and Lee also point out that even though line segments can be shared between chains, in the binary search of chains, each line segment need be accessed only once. If the test for chain B is below the test for chain A in the tree of figure 14 then any line segment on chain A which is also on chain B will only appear in the test for chain A. For example, line segment S is shared on chains 4 and 5. Since chain 5 is above chain 4 on the search tree line segment S only need be tested on chain 5. Hence for the point shown in Figure 14, it lies to the right of chain 3 since it lies to the right of

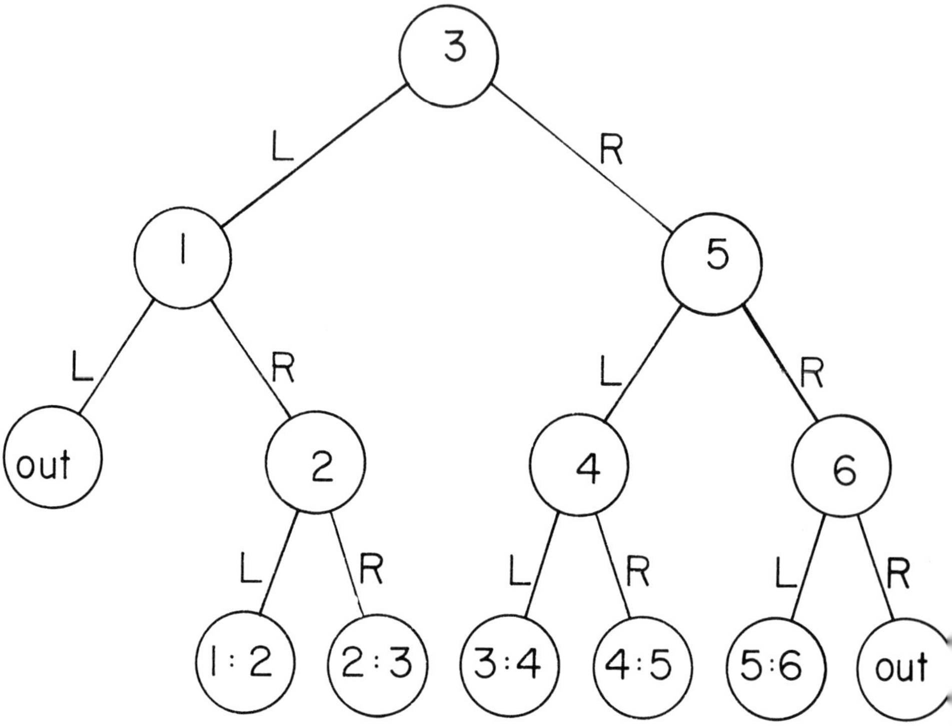

Figure 14 illustrates a binary search tree which can be used to determine between which pairs of successive chains in figure 13 a given point lies. The notation x:y means to the right of chain x and to the left of chain y.

line segment K of chain 3. And of all chains that line segment K participates in, chain 3 is the rightmost one. According to the tree, we must next test to determine on which side of chain 5 the point lies. We find that it lies to the left of chain 5 since it lies to the left of line segment S. Of all chains that line segment S participates in, chain 4 is the leftmost one. Hence the given point lies between chains 3 and 4. It is to the right of line segment K and to the left of line segment S. This information is sufficient to determine the region.

Figure 15 illustrates how the spatial data structure can store the data required for the Preparata and Lee algorithm. For the purpose of point in region problems it has two relations. The first relation is a unary relation consisting of the set of chains, each of which is a spatial data structure. The second relation is a sept-ary relation of the line segments. The components of the 7-tuples in this relation are line segment name, beginning point, ending point, region left, region right, chain left, and chain right. Chain left is the leftmost chain the line segment participates in and chain right is the rightmost chain the line segment participates in. Figure 16 illustrates the spatial data structure for a monotone chain.

V. CONCLUSION

We have described a data structure that naturally handles both raster and vector format spatial data. We have referenced the many fast vector format algorithms recently developed and have described one in detail, illustrating how our spatial data structure can fulfill its requirements. We

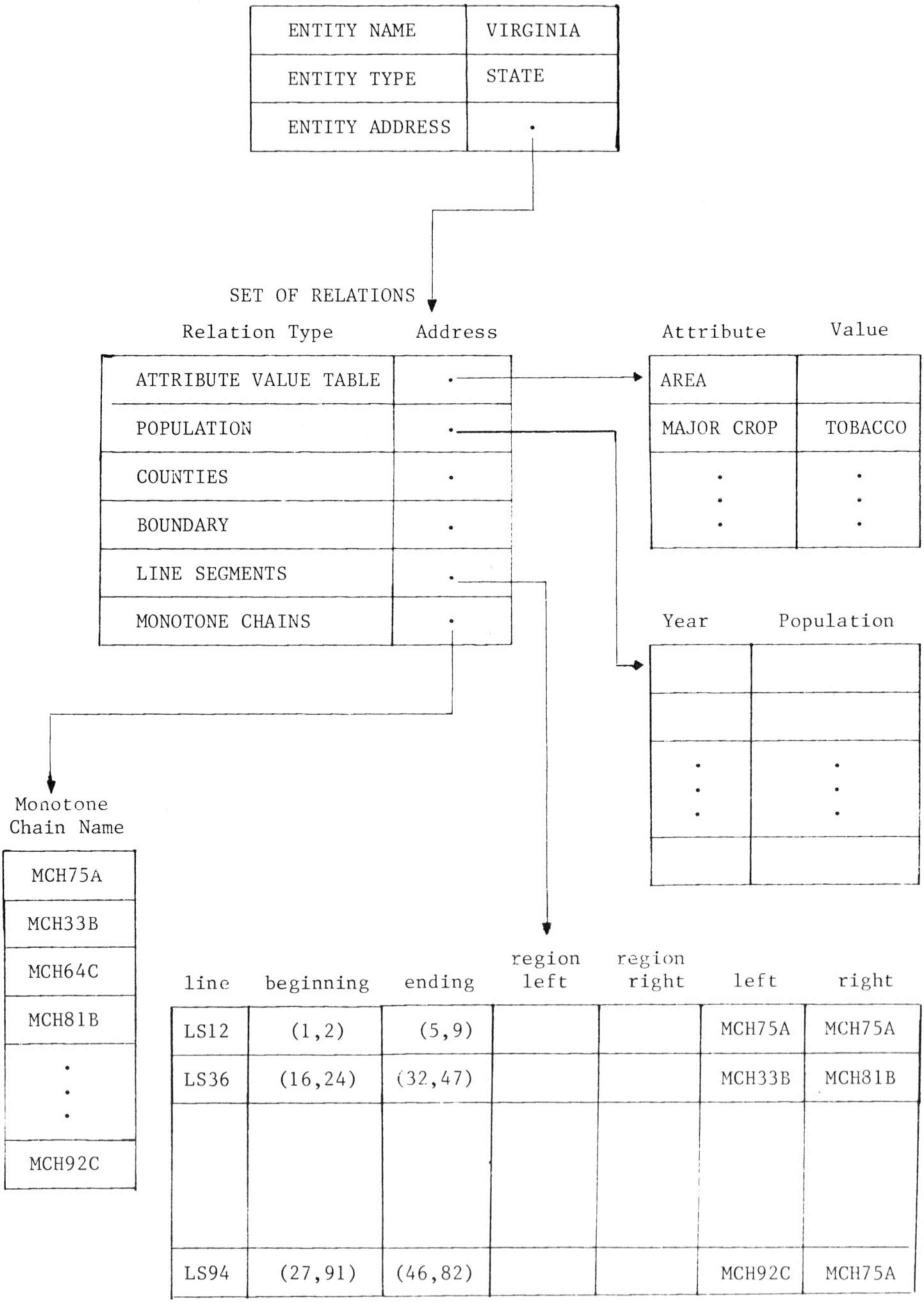

Figure 15 illustrate the spatial date structure for a region using the monotone chain algorithm for determining the region any point is ir

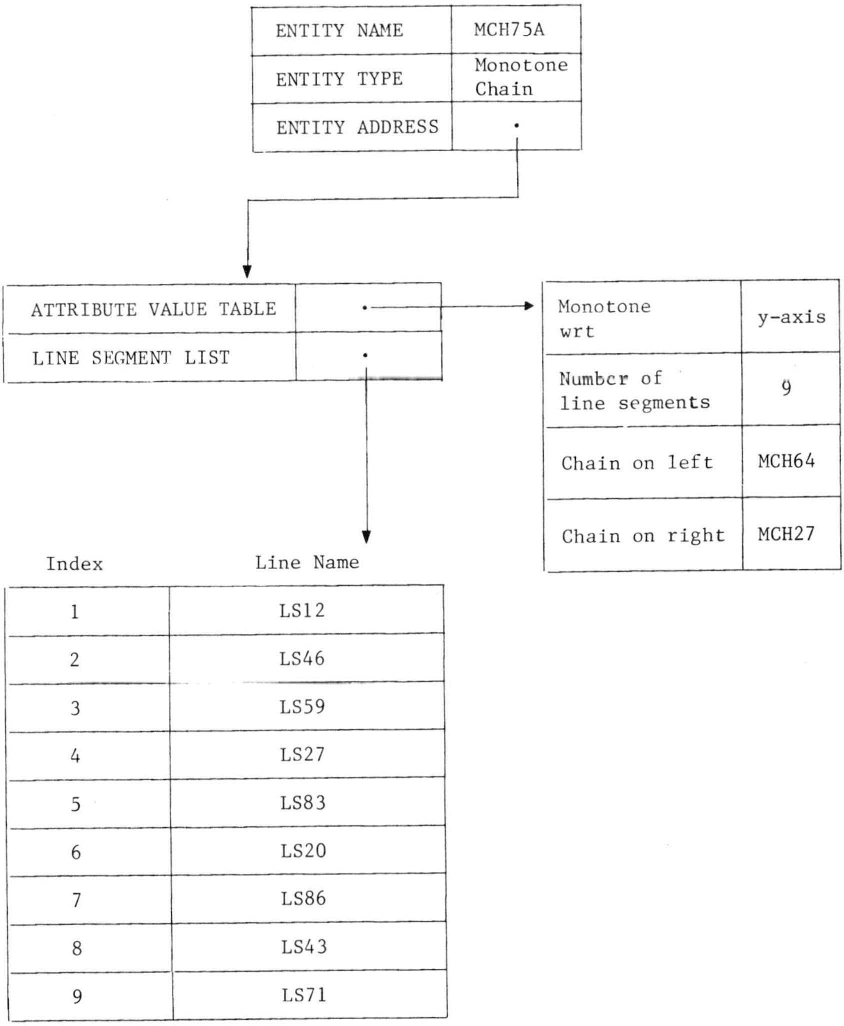

Figure 16 illustrates the spatial data structure for a monotone chain have argued that the choice of data format should not be the user's worry, but should be left up to the geographic information system which can optimize the choice of data structure based on the statistics of each particular data set and on the distribution of queries mode of that data set. We have left for another paper a description of how the geographic information can optimize computer resources by appropriate choice of data format for each particular data set.

REFERENCES

Akl, S. G. and Toussaint, G. T. (1979). "Efficient Convex Hull Algorithms for Pattern Recognition Applications," 4th IJCPR, Chicago, August 1979, 483-487.

Bentley, J. L. and Stomat, D. F. (1975). "Analysis of Range Searches in Quad Trees," Information Processing Letters, Vol. 3, No. 6, July 1975, 170-173.

Lee, D. T. and Preparata, F. P., (1977). "Location of a Point in a Planar Subdivision and Its Applications," SIAM Journal of Computing, Vol. 6, No. 3, September 1977, 594-606.

Lee, D.T. and Preparata, F. P., (1979). "An Optimal Algorithm for Finding the Kernel of a Polygon," JACM, Vol. 26, No. 3, July 1979, 415-421.

Lee, D. T. and Wong, C. K. (1977). "Multidimensional Binary Search Trees and Balanced Quad Trees," Acta Informatica, Vol. 9, 23-29.

Lee, D. T. and Preparata, F. P. (1978). "The All Nearest-Neighbor Problem for Convex Polygons," Information Processing Letters, Vol 7, No. 4, June 1978, 189-192.

Little, J. J. and Peucker, T. K. (1979). "A Recursive Procedure for Finding the Intersection of Two Digital Curves," CGIP, Vol. 10, 159-171.

Merrill, R. D. (1973). "Representation of Contours and Regions for Efficient Computer Search," CACM, Vol. 16, No. 2, February 1973, 69-82.

Nagy, G. (1978). "Advances in Computational Geometry," Proceedings of the Twenty-first Midwest Symposium on Circuits and Systems, August 14-15, 1978.

Preparata, F. P. (1979). "An Optimal Real-Time Algorithm for Planar Convex Hulls," CACM, Vol. 22, No. 7, July 1979, 402-405.

Shamos, M. I. and Hoey, D. J. (1975). "Closest-point Problems," Proceedings of the Sixteenth Symposium on Foundations of Computer Science, October 1975.

Shamos, M. I. and Bentley, J. L. (1977). "Optimal Algorithms for Structuring Geographic Data," An Advanced Study Symposium on Topological Data Structures for Geographic Information Systems, October 1977.

Shamos, M. I. (1975). "Geometric Complexity," Proceedings of the 7th Annual Symposium in Theory of Computing, May 1975.

Yang, C. C. and Lee, D. T. (1979). "A Note on All Nearest-Neighbor Problems for Convex Polygons," Information Processing Letters, Vol. 8, No. 4, April 1979, 193-194.

DESIGN OF A SPATIAL INFORMATION SYSTEM

Linda G. Shapiro

Department of Computer Science
Virginia Polytechnic Institute and State University
Blacksburg, Virginia

This paper defines a general spatial data structure for representing geographic entities and describes the use of the structure in an intelligent spatial information system. The major data structures and program control structure of the proposed system are discussed, and the operations required to answer queries are explored.

I. INTRODUCTION

Spatial data consists of geographic entities. These entities have global properties, parts, and related geographic entities. Knowledge of an entity consists of knowledge of its global properties, knowledge of its parts and their interrelationships, and knowledge of the related entities and their relationships to the given entity. In this paper we discuss the use of a formal representation structure, which we call a spatial data structure, as the building block of a spatial information system. A spatial data structure is a set of relations used to represent a geographic entity. It is a rich and flexible structure and can be used to represent any kind of spatial information from high-level entities such as cities to low-level entities such as points, lines, or pixels.

ISBN 0-12-267180-5

Geographic information systems have been important for a long time. Some of the well-known systems include the Canadian Geographic Information System (13,14), the U.S. Census DIME files (2), and the POLYVRT system (6). These systems all represent geographic data as polygons; the exact data structure used varies from system to system. Polygonal representations are also used by Edwards, Durfee and Coleman (3) who have a hierarchical data structure and by Burton (1) who stores polygons with ordered lists of points for fast searching. Go, Stonebraker, and Williams represent maps using polygons in a simple relational database.

Surface data has been represented by the triangle structure. A set of triangles represents a three-dimensional surface, each vertex containing such information as elevation and slope (5,7). The GEOGRAF system (9) was proposed to handle both planar and surface data using areas, chains of points, triangles, and high-level structures. Finally, surface or planar data can be stored in raster form (8).

We have shown (11) that the spatial data structure defined in this paper can store any of the information stored by any of the above systems and can do so without changing the logical organization of the data. In the remainder of this paper we will define the spatial data structure, give examples of its use, and describe its role in a spatial information system.

II. A SPATIAL DATA STRUCTURE

The basic kinds of data found in maps are points, lines and areas. This data can be stored in many forms; grid cells, lists of points, lists of line segments, and polygons are common examples. The best storage representation is the one that can be accessed most efficiently for a given application. Unfortunately, if everyone uses a different data structure, very little sharing of software or data can take place. In this section we define a general spatial data structure that can be used to represent any spatial or relational data.

An *atom* is a unit of data that will not be further broken down. Integers and character strings are common examples of atoms. An *attribute-value table* A/V is a set of pairs $A/V = \{(a,v) \mid a$ is an attribute and v is the value associated with attribute $a\}$. Both a and v may be atoms or more complex structures. For example, in an attribute-value table associated with a structure representing a person, the attribute AGE would have a numeric value, and the attribute MOTHER might have as its value a structure representing another person.

A *spatial data structure* D is a set $D = \{R1,\ldots,RK\}$ of relations. Each relation Rk has a *dimension* Nk and a sequence of *domain sets* $S(1,k),\ldots,S(Nk,k)$. That is for each $k = 1,\ldots,K$, $Rk \subseteq S(1,k) \times \ldots \times S(Nk,k)$. The elements of the domain sets may be atoms or spatial data structures. Since the spatial data structure is defined in terms of relations whose elements may themselves be spatial data structures, we call it a *recursive structure*. This indicates 1) that the spatial data structure is defined with a recursive definition, and 2) that it will often be possible to describe operations on the structure by simple recursive algorithms.

A spatial data structure represents a geographic entity. The entity might be as simple as a point or as complex as a whole map. An entity has global properties, component parts, and related geographic entities. Each spatial data structure will have one distinguished binary relation containing the global properties of the entity that the structure represents. The distinguished relation is an attribute-value table and will generally be referred to as the A/V relation. When a geographic entity is made up of parts, we may need to know how the parts are organized. Or, we may wish to store a list of other geographic entities that are in a particular relation to the one we are describing. Such a list is just a unary relation, and the interrelationships among parts are n-any relations.

For example, we may represent the state of Virginia by a spatial data structure. In this case, the A/V relation would contain global attributes of the state such as population, area, boundary, major crop, and so on. The values of most of these attributes (population, area, major crop) are atoms. The value of the boundary attribute is a spatial data structure defining the boundary.

One obvious division of the state is into counties. A list of counties could be included as one of the relations, or it might be more valuable to store the counties in a region adjacency relation, a binary relation associating each region (county) with every other region (county) that neighbors it. Counties, of course, would also be represented by spatial data structures.

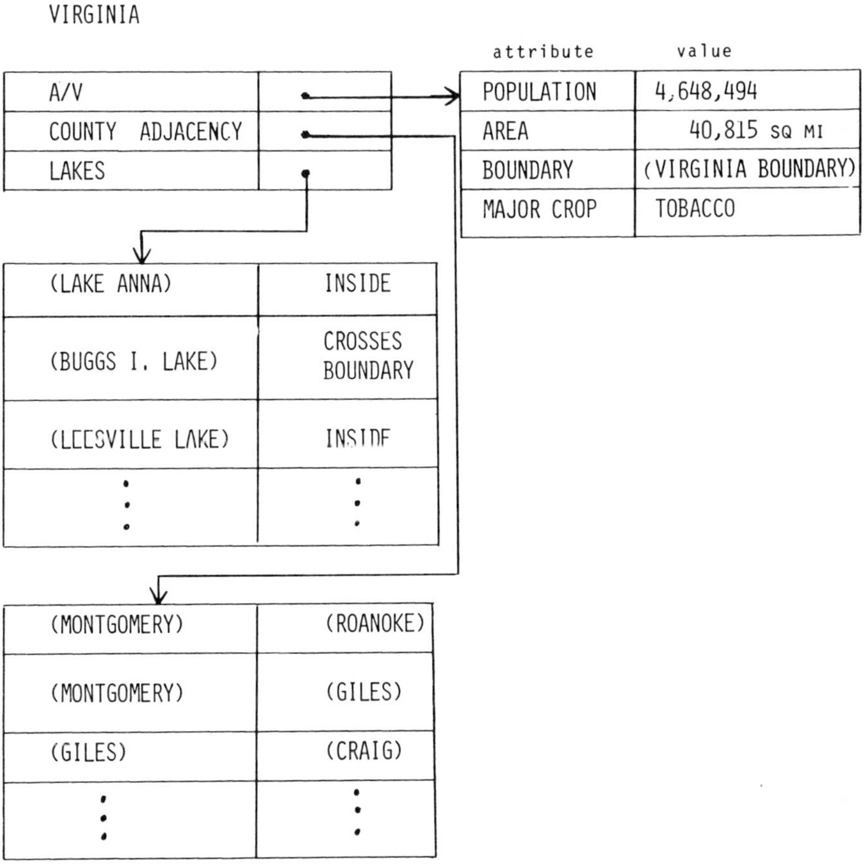

Figure 1 illustrates a spatial data structure representing the state of Virginia, The A/V relation is an attribute-value table. The COUNTY ADJACENCY relation contains pairs of adjacent counties. The LAKES relation contains pairs consisting of a lake and an indication of whether it lies inside or on the boundary of the state.

Some other geographic entities that are related to a state are its highways, railroads, lakes, rivers, and mountains. Some of these entities will be wholly contained in the state and others will cross its boundaries. One way to represent this phenomenon is to use a binary relation where the first element of each pair is a geographic entity, and the second element is a code indicating whether the entity is wholly contained in the state. The spatial data structures representing the geographic entities themselves would contain more specific information about their locations. Figure 1 illustrates a simplified spatial data structure containing an attribute-value table, a county adjacency relation, and a lakes relation for the state of Virginia.

III. THE SPATIAL INFORMATION SYSTEM

We are currently involved in the design and implementation of an experimental spatial information system using the spatial data structures concept of Section II. The system will answer user queries and solve problems presented to it in a subset of English. In this section, we describe some of the important features of the proposed system.

III.1 Major Data Structures

The spatial data structure is the primitive or building block of the system. A finite number of spatial data structure types will be allowed. For instance, the system might include spatial data structures representing the high-level entities states, cities, counties, highways, rivers, lakes, and mountains and the lower-level entities boundaries, simple boundaries, and chains. Thus the system might contain a spatial data structure whose name is MONTGOMERY and whose type is COUNTY.

For each type of spatial data structure, the system will keep a prototype structure. The prototype will indicate what attributes may be found in the A/V relation of this type of spatial data structure and what relations besides the A/V relation comprise the data structure. Similarly a finite number of relation types will be allowed, and the system will keep prototypes of the allowable relations. Thus the STATE prototype might indicate that all spatial data structures of type STATE have a COUNTY ADJACENCY relation. The COUNTY ADJACENCY prototype would indicate that this is a binary relation and that both components of each pair in the relation are spatial data structures of type COUNTY. (See Figure 3 in Section III.3.) A user query might involve a specific spatial data structure or a specific type of spatial data structure. For fast access in either case, the system will include a spatial data structure name dictionary that maps a name to a spatial data structure and a spatial data structure type dictionary that maps a type to a list of all spatial data structures of that type. Similarly a relation type dictionary will map a relation type to a list of all relations of that type. We are planning to implement relations as relational trees (10). Note that all of these structures can be represented by spatial data structures, unifying the whole system.

III.2 Program Structure

We envision an intelligent system that can handle a variety of queries. For example,

1) What is the population of Richmond?

2) What cities with population greater than 20,000 lie in Montgomery County?

3) What crops are grown within a radius of one hundred miles from Charlottesville?

4) In what counties is 20 per cent of the area mountainous?

5) What is the shortest route from Alexandria to Blacksburg?

are all reasonable queries. Some of the questions require quick table-look-ups to produce the answer and some require searching through the database and using special purpose algorithms. It is also possible that there are several different procedures or several different paths through the database to find an answer to a query. In this case, the system should be able to choose the most efficient way to solve the problem. We propose the following program organization.

The user queries are processed by a PARSER routine which translates the query into a component command and a list of qualifiers. The PARSER will detect any syntax errors and try to rectify the mistakes through dialog with the user. The output of the PARSER is accepted by the TREE-GENERATION routine which creates a tree of possible solution paths through the database. The most efficient solution path is determined by the TREE-EVALUATION routine which produces, as its output, a procedure of primitive database functions which when executed will satisfy the user query. Finally, the EXECUTOR executes the procedure to complete the command. Figure 2 depicts the flow of a user query from input to answer.

Each of the major modules must access information structures as part of its input. The PARSER requires a grammar that describes the syntax of the query language. Since the query language should be English-like, the PARSER will also need some world knowledge to help disambiguate some of the phrases used. The TREE-GENERATION routine uses world knowledge, spatial knowledge, and knowledge of the database

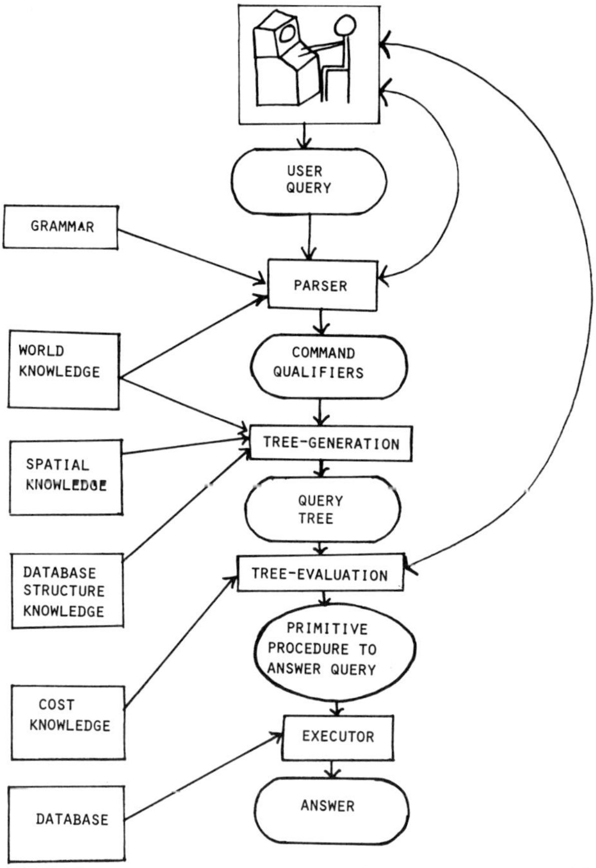

Figure 2 illustrates the flow of a user query through the spatial information system.

structure. The world knowledge and spatial knowledge can be represented in several different ways including relational tables, production rules, and procedures. Again each of these knowledge representations can be represented by spatial data structures. The database structure knowledge should include the spatial data structure name dictionary, the spatial data structure type dictionary, the relation type dictionary, and all the prototypes.

The TREE-EVALUATION routine uses cost knowledge to choose a path from the set of paths produced by the TREE-GENERATION routine. The cost knowledge may be stored in tabular form or built into the TREE-EVALUATION routine in the form of procedures. The TREE-EVALUATION routine may be offered a choice between a high cost for an exact answer or a lower cost for a less than perfect answer. The input to the EXECUTOR is a procedure of primitive functions that operate on the database. The EXECUTOR has access to the entire database of information. The EXECUTOR can also access the primitive functions and special purpose high-level algorithms that are built into the system.

III.3 Primitive Operations

The spatial data structure is a relational structure. Because of this, the spatial database system shares many characteristics of relational database systems. In particular, all of the primitive operations used in relational database systems are applicable to the relations of a spatial data structure. We will use a small example database to motivate the use of these and other primitive operations.

Figure 3 illustrates a set of prototypes for spatial data structures and their relations that might be found in a spatial information system. The STATE prototype indicates that STATE is a type of spatial data structure having an A/V relation, a COUNTY ADJACENCY relation, and a RIVERS relation. The A/V relation has four attributes: NAME, whose value is a character string, POPULATION and AREA whose values are numbers, and BOUNDARY whose value is a spatial data structure of type POLYGON.

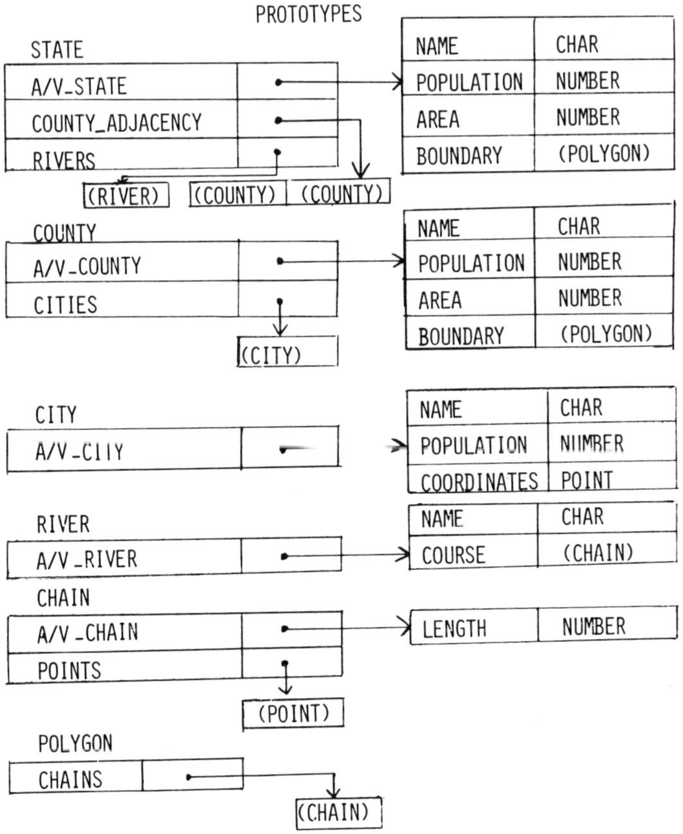

Figure 3 illustrates a set of prototypes for the spatial data structure types STATE, COUNTY, CITY, RIVER, CHAIN, and POLYGON and the relation types COUNTY ADJACENCY, RIVERS, POINTS, and CHAINS.

The COUNTY ADJACENCY relation is a binary relation, and each member of each pair is a spatial data structure of type COUNTY. The RIVERS relation is a unary relation, and each element is a spatial data structure of type RIVER. The other prototypes convey similar information.

The following questions are possible queries to a spatial information system having the prototypes of Figure 3. Under each question, we suggest a sequence of operations that might be performed to answer the query.

```
1)  What cities are in county X?

    A.  Locate county X.
    B.  For each CITY C in CITIES(X).
        1.  Look up N = NAME(C).
        2.  Add N to the relation being created.


2)  What cities are in state X?

    A.  Locate state X.
    B.  Perform a projection operation on COUNTY
        ADJACENCY(X) to obtain a list of
        counties.
    C.  For each county Y in the list
        For each city C in CITIES(Y)
            1.  Look up N = NAME(C).
            2.  Add N to the relation being created.


3)  What is the length of river R?

    A.  Locate river R.
    B.  Return value of LENGTH(COURSE(R)).


4)  What cities lie on rivers in state X?

    A.  Locate state X.
    B.  Perform a projection operation on COUNTY
        ADJACENCY(X) to obtain a list of
        counties
    C.  For each county Y in the list
        For each city C in CITIES(Y)
            For each river R in RIVERS(X)
            if POINT_CHAIN_DISTANCE(COORDINATES(C),
            COURSE(R))=Ø
            then add C to the relation being created.


5)  What cities are within 15 miles of a river in state
    X?

    same as 4) except change '=Ø' to '15'.


6)  What counties in state X does river R flow through?

    A.  Locate state X.
    B.  Perform a projection operation on COUNTY
        ADJACENCY(X) to obtain a list of
        counties.
    C.  For each county Y in the list
        if CHAIN_INTERSECTS_POLYGON(COURSE(R), BOUN-
        DARY(Y))
        then add Y to the relation being created.
```

7) What states does river R flow in?

   For every state S
       if R is an element of RIVERS(S) then add S to
       the relation being created.

8) What points do river X and river Y share?

   A. Locate river X.
   B. Locate river Y.
   C. Construct the intersection of POINTS(COURSE(X)) with POINTS(COURSE(Y)).

9) Answer the following questions about the region R defined by the states X,Y, and Z.

   A. Locate state X.
   D. Locate state Y.
   C. Locate state Z.
   D. Create a new temporary spatial data structure R from the information in X,Y, and Z and dependent on the prototype for spatial data structures of type REGION. This will include such operations as finding sums of populations and areas and constructing a new boundary based on three existant boundaries.

From these sample queries, we find the following operations are necessary.

1. projection in the relational database sense
2. selection in the relational database sense
3. intersection or join in the relational database sense
4. look up the value of an attribute
5. call on geometric or distance functions
6. create a list
7. add elements to a list
8. comparison
9. determine if an N-tuple is a member of a relation
10. create a new relation
11. create a new spatial data structure

III.4 Semantics

The spatial information system must translate a natural language query or command to a set of primitive operations to be performed on the database. We have not yet specified how the semantics or meaning of a query is to be handled. For example, consider query 6), "What counties in state X does river R flow through?". The parser will break up the query into an intermediate form such as

```
FIND  {C | C is a county
           C is in state X
           River R flows through C}
```

With respect to the database, "C is a county" means "TYPE(C) = COUNTY", "C is in state X" means "C ∈ PROJECTION (COUNTY_ADJACENCY(X))", and "River R flows through C" means "CHAIN_INTERSECTS_POLYGON (COURSE(R),BOUNDARY(C)) = true" where PROJECTION is a function that implements the projection operation and CHAIN INTERSECTS POLYGON is a function that determines if a chain of points intersects a polygon.

One method of translating the English phrases to primitive operations is via a production system (12). A production system keeps its knowledge as a set of productions, each of the form "ANTECEDENTS → CONSEQUENTS". ANTECEDENTS is a list of conditions that must be true for the production to be applied, and CONSEQUENTS is a list of facts that become true after the production is applied. Thus the following three productions could be used to represent the above knowledge.

```
C is a county      ->  TYPE(C) = COUNTY

C is a county      ->  C ∈ PROJECTION
C is in state X        ((COUNTY ADJACENCY(X))

R is a river       ->  R ∈ RIVERS(X)
R is in state X
```

IV. DISCUSSION AND CONCLUSIONS

We have defined a spatial data structure to be used as a building block in a spatial information system. The database would consist of a set of spatial data structures, each consisting of a set of relations and each representing a geographic entity. The proposed system will contain a control processor which, when given a query, will determine all possible paths through the structure to answer the query and select the best path with the use of a cost function. The use of production systems will add semantic capabilities to the system.

The proposed spatial information system differs from former systems in its unified approach to representing geographic entities. It has in common with current relational database systems the use of N-ary relations to store data. It differs from such systems in two important ways. 1) Relational database systems are normalized; that is, components of N-tuples of relations are atomic. It is part of the philosophy of our system that components of N-tuples of relations can be themselves complex structures. 2) Relational database systems are for storage and retrieval, only. The proposed system will be more intelligent; it will also have a deductive capability. It will be able to translate natural language input and to determine the best method for answering a query. We expect the system to be a useful aid for spatial analysis.

REFERENCES

1. Burton, W., "Representation of Many-Sided Polygons and Polygonal Lines for Rapid Processing," Communications of the ACM, Vol. 20, No. 3, March 1977, pp. 166-170.

2. Cook, D. and W. Maxfield, "The Development of a Geographic Base File and Its Uses for Mapping," Proceedings of URISA, Garden City, Long Island, September 1967.

3. Edwards, R.L., R. Durfee, and P. Coleman, "Definition of a Hierarchical Polygonal Data Structure and the Associated Conversion of a Georgraphic Base File from Boundary Segment Format," An Advanced Study Symposium on Topological Data Structure for Geographic Information Systems, Harvard University, Cambridge, Massachusetts, October 1977.

4. Go, A., Stonebraker, M., and Williams, C., An Approach to Implementing a Geo-Data System, Memo No. ERL-M529, Electronics Research Laboratory, College of Engineering, University of California at Berkeley, 1975.

5. Gold, C., "Triangular Element Data Structures," Users Applications Symposium Proceedings, The University of Alberta Computing Services, Edmonton, Alberta, Canada, 1976.

6. Laboratory for Computer Graphics, "POLYVRT: A Program to Convert Geographic Base Files," Harvard University, Cambridge, Massachusetts, 1974.

7. Males, R., "ADAPT - A Spatial Data Structure for Use with Planning and Design Models," An Advanced Study Symposium of Topological Data Structures for Geographic Information Systems, Harvard University, Cambridge, Massachusetts, October 1977.

8. Peuquet, D.J., "A Raster-Mode Algorithm for Interactive Modification of Line Drawing Data", Computer Graphics and Image Processing, Vol. 10, 1979, pp. 142-158.

9. Peucker, T.K. and N. Chrisman, "Cartographic Data Structures", The American Cartographer, Vol. 2, No. 1, April 1975, pp. 55-69.

10. Shapiro, L. G., "Data Structures for Picture Processing" to appear in Computer Graphics and Image Processing, 1979.

11. Shapiro, L. G. and R. M. Haralick, "A General Spatial Data Structure", Proceedings of the IEEE Conference on Pattern Recognition and Image Processing, May 31-June 2, 1978, Chicago, pp. 238-249.

12. Shortcliffe, E. H., Computer-Based Medical Consultations, MYCIN, Elsevior, New York, 1976.

13. Switzer, W. A., "The Canada Geographic Information Sy Stem", Automation in Cartography, eds. J. M. Wilford-Brickwood, R. Bertrand, and L. van Zuylen, International Cartographic Association, The Netherlands, 1975.

14. Tomlinson, R., "A Geographic Information System for Regional Planning," Land Evaluation (Stewart, ed.), McMillian of Australia, Sydney, Australia, 1968.

WHAT IS A "GOOD" DATA STRUCTURE
FOR 2-D POINTS?

George Nagy

Department of Computer Science
University of Nebraska-Lincoln
Lincoln, Nebraska

This note poses the problem of finding the nearest neighbor of a search point among a given set of points in the plane under the condition that the set must be updated from time to time. Sundry unsuccessful encounters with various aspects of this problem in a geographic data processing environment are traced, and some relevant results from the area of computational geometry and from other sources are summarized.

INTRODUCTION

The problem discussed in this edited and padded version of an informal presentation at the NATO Advanced Study Institute in Maratea is recommended for diversion during salvable time intervals as a substitute for crosswords, illustrated magazines, and other less edifying pastimes. Because I have been devoting lost moments to this problem for more than five years with great enjoyment but very little progress, I welcomed the opportunity to share it with the distinguished and amiable participants of the Institute, and am also pleased to present it now to a wider audience.

Problem Statement

The problem consists of finding a data structure (and associated update and query algorithms) for storing a set of two-dimensional points under the following conditions:

ISBN 0-12-267180-5

1) The precision of the locational coordinates of the points is high compared to the total number of stored points. In other words, the number of possible point-locations is large compared to the actual number of points: the extent under consideration is sparsely populated.

2) The extent under consideration is limited. The minimum and maximum values of the coordinates are given.

3) The data structure must allow efficient implementation of queries of the following type: What is the nearest point in the current set (or, perhaps, what is a "close" point) to a new point called the search point?

4) The data structure must allow efficient updating of the stored set of points through the addition of new points. The ratio of queries to updates is fixed. (This aspect is reminescent of the problem of the optimum location of copies of files in a distributed database with a given network configuration and fixed transmission costs (1).)

5) A worst-case solution, independent of the actual configuration of the stored points and of the search points, is sought. However, a solution with good statistical characteristics, based perhaps on a uniform distribution of points in a rectangle or a circle, would also be of interest.

This completes the formulation of the problem. The remainder of the Introduction consists of miscellaneous items which may or may not motivate the reader to tackle the task. Such additional clues as are within the author's ken are provided.

Historical Note

A few words may be in order to explain any prejudices or predilections which may surface. Having made the transition from an early and idealistic interest in brain models to perception, pattern recognition, optical character recognition and, eventually, terrain classification by remote sensing methods, in 1970 it occurred to me that even if we succeeded in classifying every single one of the billions (yes) of pixels about to be generated daily by the Earth Resources Technology Satellites (ERTS), we could do little with this information except plot it in the form of thematic maps. My interest thereupon shifted to geographic data processing systems, which were then in their infancy. (At their current stage of development, most geographic data processing systems, like small children, are still more trouble than they are worth, but some are interesting to interact with and, once in a rare while, produce something worthwhile.)

In 1974, having read Knuth's Volume III (2) and sat in on some lectures in the sophomore data-structures course, I began to wonder about the applicability of these fancy concepts - queues, deques, stacks, pointer structures - to geographic data processing. Since at the time we were working with the locations of center-pivot irrigation rigs in Nebraska, I was particularly interested in the use of such data structures for inventorying two-dimensional point data. At the 1975 Detroit Computer Science Conference I had the good fortune to run into Professor Lee White of Ohio State University, a noted authority on data structures. I attempted to interest him in the problem. After mulling it over, he assigned it to his graduate seminar where one of the students, Chin San Wang (now with Univac), adopted it and produced the contributions mentioned below.

Shortly thereafter Ian Shamos published a series of important and beautiful papers securing the foundations of the new field of computational geometry (3,4,5). Attracting the attention of several outstanding computer scientists, (to mention only two: Jon Bentley (6) and Franco Preparata (7)), the field burgeoned, and a number of results closely related to "my" problem were obtained.

From 1976 to 1978 I worked with a graduate student (Sharad Wagle, now Assistant Professor at the Illinois Institute of Technology) on several geographic data processing problems of local interest. By way of respite from irritating hardware failures and unaccountable programming mistakes, we often discussed the point-set data-structures problem, but I can no longer recall which false leads he thought of and which were the product of my own misguided imagination.

At the NATO Institute itself, on a dark night on the terrace, one of the participants, Dr. Bernd Schulze, made an interesting suggestion which is also reported below for the sake of completeness.

A frustrating characteristic which our problem shares with many others in computational geometry, pattern recognition, and image processing, is that human beings can solve by inspection, virtually instantaneously, most of the problems associated with two-dimensional point sets, even with a large number of samples! Consequently, in lieu of solving the problem, at this stage we would be quite satisfied with a proof that a good algorithmic solution does not exist. This would legitimize some form of interactive operation, and allow us to go on to other things.

Point Sets in Geographic Data Processing

Examples of point sets of interest in GDP are irrigation wells (there are currently over 60,000 registered in Nebraska: for inventory purposes their locations must be specified to within a quarter-section), cities and towns on small-scale maps, benchmarks for geodetic and cadastral surveys, and the locations of schools, fire stations, hospitals, and other "service centers" on urban maps. A relatively new example with some interesting properties is the location of MX missile sites.

Questions that one might ask about geographic points include the following:

1) Is a given point already included among the existing points? (This is the <u>point-identity</u> problem.)

2) To what point among the existing points is a new point closest? (The <u>nearest-neighbor</u> problem.)

3) Which two points among the existing points are closest together? (The <u>two-nearest neighbors</u> problem.)

4) What is the optimal sequence of traversing all the points according to a given criterion? (The <u>point-ordering</u> problem.)

5) In which one of a set of non-overlapping polygons does a given point lie? (The <u>point-inclusion</u> problem.)

6) What locations represent cluster-centers among the given points? (The <u>generalization</u> problem.)

Conventional Storage Methods for Point-Sets

The type of question we are likely to ask about a point set naturally influences the manner in which we would store it in a computer. Other important factors include the total number of points that we must deal with, the relative cost of the available storage devices, and the frequency with which we expect to update the list.

Among the widely-used methods available to store point sets are the following:

1. Matrix Storage. Here the extent under consideration is divided ahead of time into a fixed number of binary cells. A "1" in a cell signifies the presence of a point, and a "0" the absence. This type of data structure can be updated very rapidly, but the resolution with which the coordinates of any given point are preserved is limited by the number of rows and columns in the matrix. The amount of storage space required is fixed regardless of the number of points in the set.

2. Sparse Matrix Storage. Most sparse-matrix storage schemes are based on storing the relative displacement of the entries in the original matrix. Sophisticated data-compression methods for large binary matrices can be found in the literature of facsimile devices and black-and-white image compression (8,9,10). Such methods are based on one and two-dimensional run-length codes, adaptive delta modulation, and on predictive codes. The accent in all these methods is on efficient data storage; since they are concerned with fixed images, the question of updating the point set does not arise. The amount of storage required is a function of the total number of points and of their disposition on the underlying two-dimensional grid.

3. List of Vectors. Here the x and y coordinates of each point are stored in two consecutive storage locations, with the precision governed by the word-length of the

supporting hardware and software. The coordinates may be sorted with the x-coordinates serving as the primary key and the y-coordinates as the secondary key, or vice-versa. Alternatively, they might be sorted first on the basis of distance from an arbitrary origin, then by angular position. None of these sorting orders provides a really satisfactory search procedure for distance-related queries. The storage required is proportional to the number of points.

4. Display or Plot List. If the points are to be plotted on a raster plotter (such as a Gould electrostatic plotter), then the matrix method is quite satisfactory, with the number of columns in the matrix set equal to the number of nibs in the plotter. If, however, the points are to be plotted on flat-bed incremental plotter, then the fastest plotting time will be achieved (to a first approximation, without considering pen acceleration) by storing and plotting the points in the order which minimizes the length of the path connecting them. This is, of course, the classical traveling-salesman problem. From recent advances in computational complexity, we know better than to try to find the absolutely best solution: we are wiser to accept a "good" solution, and, if necessary, buy a faster plotter. Similar considerations, with a reduction in the scale of the time dimension of perhaps a factor of 1000, apply to steered-beam display devices (11). The storage is again proportional to the number of points, but some saving can be achieved by storing only the displacement of each point relative to its predecessor.

Computational Geometry and Geography

The problem posed above falls squarely in the realm of computational geometry. Computational geometry is the study of the efficiency of data processing operations on geometric entities such as points, curves, and polygons (3). Many of its applications in geography bear on large-scale retrieval systems where the geographical information is stored in computer-readable form and extensive computations are required to update the database or to retrieve items satisfying a specific constraint (12-16). In such systems the transformations in data format between available data sources (such as coordinate digitizers or satellite tapes), storage media (for instance, magnetic disk),and output devices (graphic displays or incremental plotters) also require efficient algorithms (17,18).

What is meant, then, by "efficient" algorithms? One aspect of efficiency is the number of computer operations required to perform a designated function, such as responding to a query or producing a specified map. The number of computer operations or the computational effort required depends on the parameters of the problem being solved. Typically, one of those parameters dominates all others in determining the computational effort. As an example, the number of regions in a thematic map is likely to have paramount influence on the computational effort required to answer certain map-related questions.

In all algorithms, the computational effort grows with 'n', the value of the dominant parameter. The growth can be linear, quadratic, etc., in 'n'. The rise in the computational effort as a function of 'n' is called the computational complexity of the algorithm (19).

Storage complexity is similarly defined as the rate of growth in the storage required for the data as a function of the problem parameter 'n'. Thus, for an algorithm with storage complexity proportional to n^2, the storage requirements quadruple for every doubling of 'n'. By defining complexity measures as rate of growth instead of in absolute terms, the complexity-analysis of algorithms avoids complications arising from variations in the operational cycle time of computers, in programming languages and in storage media (all of which have only relatively minor effect on the performance of the algorithm).

A third factor related to algorithmic efficiency is the need for preprocessing the data to convert it to a form suitable for computation. Preprocessing operations are normally performed only when the data base is established or updated, hence their cost is amortized over the lifetime of the data.

In the examples below, it will be seen that logarithmic optimal solutions to geometric problems are fairly common, just as they are in one-dimensional sorting and search problems. In both cases, the logarithmic solution is a simple consequence of the fact that the number of points eliminated from consideration at each step is proportional to the remaining number of points.

Not taken into account at all in computational geometry is the cost of developing and maintaining the necessary software in a particular computer environment. In many instances this cost is the dominant consideration in choosing among alternative means of implementing a system. Computational geometry may thus provide only an idealized solution to problems, and is only one tool among many to guide the designer in developing an acceptable system.

By way of illustration of current research in computational geometry, two recently discovered methods to solve the problem of point inclusion are described. It will also be shown that the nearest-neighbor problem can be approached in terms of point inclusion.

Point Inclusion

The determination of whether a point is contained within a region is one of the more common requirements in geographic data retrieval. It arises, for instance, in locating a point (specified by its coordinates) with respect to city, county, or district boundaries.

The slab method, due to Shamos (5),provides one efficient method of determining in which (convex) region of a polygonal map a point lies.

The notion of a binary search is central to the implementation of this, and many other, algorithms. The essence of the concept is that, given any ordered (sorted) set of n scalar values, one may determine where a new value (i.e., the search value) belongs in the set with a number of comparisons equal to $\log_2 n$.

In Shamos' point-inclusion method, the vertices of the map are first sorted according to their vertical (y) coordinates, which are considered as dividing the plane into horizontal slabs. Within any slab, the boundary segments do not cross, and hence, can be sorted in left-to-right order.

These operations are independent of the search point, and are carried out when the polygon map is first made available. They do not affect the running time of the search algorithm and are part of preprocessing the data.

Once the data is sorted in this form, two binary searches - one among the slabs and the other within the slab found to contain the given point - suffice to determine the containing region. For n vertices, the length of the search is proportional to $\log_2 n$, with storage requirements (for up to n boundary segments within each of up to n slabs) proportional to n^2. We note in passing that even for as few as 1000 vertices, the difference between n and $\log_2 n$ is a factor of 100.

Although Shamos' algorithm is restricted to convex polygons, any non-convex polygon may be converted into a set of convex polygons by partitioning it into triangles.

Another efficient algorithm for point inclusion, devised by Lee and Preparata (20), starts by examining each vertex to determine whether it has at least one ascending and one descending edge incident upon it. If that is not the case, the polygon map (or <u>planar straight-line graph</u>) is <u>regularized</u> by introducing additional edges as necessary. The regularization of a map with n vertices can be performed with $n \log_2 n$ operations.

The next step is to assign each edge to one or more <u>chains</u> in such a way that in each chain successive vertices are ordered according to their vertical coordinates. The chains are numbered left to right, 1 to m.

The left-to-right ordering of the chains allows a binary search to locate the search point between two adjacent chains.

Since a chain cannot "double back" on itself, the edge in each chain which is cut by a horizontal line through the search point may be located with $\log_2 s$ comparisons by means of a second binary search among the s vertices of the chain. After this edge is located, it is a simple matter to determine whether the search point is to the right or left of the chain.

Because the two searches are _nested_, the overall computation is proportional to $(\log_2 n)^2$, versus $\log_2 n$ for Shamos' method, but it requires storage proportional only to n. The choice between the two algorithms for point inclusion described in this section thus depends on the relative cost of storage and of computation, on the frequency of searches in comparison to how often the database is updated, and on the total number of nodes and edges in the polygon map under consideration.

Nearest Neighbor

The efficient determination of the _nearest neighbor_ of a specified point among a set of predetermined points is based on the computation of the _proximal polygon_ corresponding to each point in the set. The edges of these polygons are the right bisectors of the straight lines joining pairs of points. They are called _Voronoi polygons_ in geometry and _Thiessen polygons_ in geography. The Thiessen polygons subdivide the plane into mutually exclusive regions. To find the nearest neighbor of a search point, it is sufficient to determine which of these polygons contains the specified point. The nearest-neighbor problem is thus transformed into a point-inclusion problem. Because each region is _convex_, Shamos' algorithm is directly applicable and the regularization step in Preparata's and Lee's algorithm may be omitted. It can be shown that the Thiessen polygons for n points may be computed with a number of operations proportional to n $\log_2$ n (3).

A common geographic application of the nearest neighbor problem occurs in facilities location, as for instance in determining the closest supply of energy to each prospective plant site (21,22).

A number of other results in computational geometry of potential interest to geographers are discussed in (23).

2-D Pointer Structures

My original approach to the problem was based on an attempt at a two-dimensional generalization of a doubly-threaded linked list. Each of the stored points would have four pointers associated with it, with each pointing to the point's nearest neighbor in one of the four quadrants. This structure does not, however, preserve the important reciprocal property of the doubly-linked list: if a point A is the nearest northeast neighbor of a point B, it does not follow that B is the nearest southwest neighbor of A.

Professor White described the problem to his class as follows:

CIS 680 Data Structures - Problem #11.

"At the 1975 Computer Science Conference, George Nagy challenged me with the following problem. Given a set of N points in the plane, assume there are presently four pointers Pnw(x), Pne(x), Pse (x), Psw (x) associated with each point x. Consider the 4 quadrants formed by choosing each point x as the origin. We are interested in the nearest points to x in each quadrant, and these are indicated by the 4 pointers indicated above. If no point exists in a specific quadrant of x, then the respective pointer is set null. The problem is to superimpose additional data structures on this set of points so that points may be inserted or deleted more easily without performing calculations involving all N points and updating all pointers. The problem is especially severe for insertion, where the challenge is to update all pointers with a minimum for processing, which does not depend upon the number of points N currently present."

Chin Wang introduced two additional constructs, the "z-neighborhood" and the "terminal bank" (24).

The northeast z-neighborhood of a point Z consists of all the points W such that W is in the northeast quadrant of Z but the nearest southwest neighbor of W is not in the northeast quadrant of Z. The three other z-neighborhoods are defined analogously. It can be shown that if a new point is inserted at Z, pointers associated with points outside the Z-neighborhood of Z need not be updated. An "improvement lemma" can be used to further reduce the number of points in the northeast Z-neighborhood of Z by eliminating all points in the northeast quadrants of other points in that Z-neighborhood.

The northeast terminal bank is defined as the set of all roots in the forest of trees constituted by all southwest pointers. Wang showed that all four types of Z-neighborhoods can be efficiently searched starting only with roots in the northeast terminal bank (or in any one of the other three banks). Algorithms for updating the terminal banks were also demonstrated.

Wang and his associate John Yang conjectured that when the points are randomly distributed the size of the terminal banks and of the individual Z-neighborhoods, and the number of comparisons necessary for insertion, grow only logarithmically with the total number of points. These conjectures were based on intuitive arguments and on simulation results but, even if true, they do not provide a complete answer to our problem.

Another idea, suggested as mentionned above by Dr. B. Schulze of Mûnster University, is to establish a triangular network linking all the points. Several triangulation algorithms yielding "compact" triangles are available in the literature. Their most common application in geography is for contouring, by linear interpolation, data based on irregularly spaced points.

The data structure would consist of three pointers linking each triangle with the three adjacent triangles. Such a network is easy to update by subdividing the triangle in which a new point falls. Only the points associated with this triangle and its three neighbors need to be changed. Unfortunately, even if in the original triangulation nearest neighbors always constitute vertices of the same triangle, the above updating algorithm does not preserve this property.

Envoy

This completes the Introduction to the point-set data-structures problem. One day I hope to be in a position to complete the remaining sections. At the Maratea Institute I offered a copy of this volume for the first successful solution, but it appears that an even more attractive prize will have to be staked.

REFERENCES

1. Casey, R.G., "Allocation of Copies of a File in an Information Network", Proc. Spring Joint Computer Conference, 617-625, AFIPS Press, New York, 1972.
2. Knuth, D.E., The Art of Computer Programming, Vol. 3, Sorting and Searching, Addison-Wesley, Reading, Massachusetts, 1973.
3. Shamos, M.I., Computational Geometry, Ph.D. Dissertation, Yale University, New Haven, Connecticut, 1978.
4. Bentley, J.L., and M.I. Shamos, "Divide-and-Conquer in Multidimensional Space", Proc. 8th Annual Symposium on Theory of Computation, pp. 220-230, ACM,1976.
5. Shamos, M.I., "Geometric Complexity", Proc. 7th Annual Symposium on Theory of Computation, pp. 224-233, ACM, 1975.
6. Bentley, J.L., "Multidimensional Divide-and-Conquer", Computer Science Research Review, Carnegie-Mellon University, Pittsburgh, Pennsylvania, 1978.
7. Preparata, F.P., (ed.) Steps into Computational Geometry, CSL Report R-760, 1977, and Steps into Computational Geometry Notebook II, CSL Report R-792, U. Illinois, Urbana, September, 1977.
8. Golomb, S.W., "Run-Length Encoding", IEEE Trans. Information Theory, Vol. IT-12, pp. 399-401, July 1966.
9. Bahl, L.R., and H. Kobayashi, Image Data Compression by Predictive Coding," IBM J. Res. Develop., Vol. 18 pp. 164-179, March 1974.
10. Simon, J.C. and A. Rosenfeld, Proceedings of a NATO Advanced Study Institute, Bonas, France, Noordhoff, Leyden, 1976.
11. Newman, W.M. and R.F. Sproull, Principles of Interactive Graphics, Second Edition, McGraw-Hill, New York 1979.
12. Nagy, G. and S. Wagle, "Geographic Data Processing", ACM Computing Surveys, Vol. 11, No. 2, June, 1979.
13. Calkins, H.W. and R.F. Tomlinson, Geographic Information Systems, Methods, and Equipment for Land Use Planning, IGU, 1977.
14. Rhind, D.W., "Computer Aided Cartography", Trans. Institute of British Geographers, New Series, Vol. 2, No. 1, 1977.
15. Taylor, D.R.F. (ed.), Proc. of Symposium on Geographic Information Processing, Dept. of Geography, Carleton U., Ottawa, 1976.
16. Tomlinson, R.F. (ed.), Computer Handling of Geographic Data, UNESCO Press, 1976.
17. Peucker, T.K., & Chrisman, N., "Cartographic Data Structures", The American Cartographer 2, #1, pp. 55-69, April 1975.
18. Peuquet, D., "A Raster-Mode Algorithm for Interactive Modification of Line Drawing Data", Computer Graphics and Image Processing 10, 142-158, 1979.
19. Aho, A.V., J.E. Hopcroft, and J.D. Ullman, The Design and Analysis of Computer Algorithms, Addison-Wesley, Reading, Massachusetts, 1974.

20. Lee, D.T., and F.P. Preparata, "Location of a Point in a Planar Subdivision and its Application", Proc. 8th Annual Symposium on Theory of Computation, pp. 231-235, ACM, 1976.
21. Nordbeck, S., and B. Rystedt, Computer Cartography, Student-litteratur, Lund, Sweden 1972.
22. Thierauf, R., and R. Klekamp, Decision Making Through Operations Research, Wiley, New York 1975.
23. Nagy, G., "Advances in Computational Geometry", Proc. Twenty-first Midwest Symposium on Circuits and Systems, pp. 598-603, Ames, Iowa, April 1978.
24. Wang, Chin San, "A Solution to G. Nagy's Geographic Data Base Problem", unpublished manuscript, presented at the IEEE Computer Society Workshop on Data Structures and Pattern Recognition, Albuquerque, New Mexico, February 1976.

TREE STRUCTURES FOR REGION REPRESENTATION

Azriel Rosenfeld

Computer Vision Laboratory
University of Maryland
College Park, Maryland

This paper describes a method of representing arrays by trees of degree 4, which are constructed by recursively subdividing the array into blocks of constant value. In particular, this method can be applied to binary arrays representing regions (1's are region points). Algorithms for conversion between this and other region representations, and for measuring geometric properties of regions represented in this way, are informally described.

1. INTRODUCTION

Region representation plays a key role in image and scene analysis, computer cartography, and computer graphics. There are a variety of approaches to representing regions, based on their boundaries or their "skeletons"; some of these are reviewed in the following paragraphs. Recently, a tree representation has been proposed which offers a number of advantages; it is also described below.

We assume in what follows that a region is an arbitrary

The support of the Defense Advanced Research Projects Agency and the U.S. Army Night Vision Laboratory under Contract DAAG-53-76C-0138 (DARPA Order 3206) is gratefully acknowledged, as is the help of Kathryn Riley in preparing this paper. The material in this paper is based on a series of reports by H. Samet, as well as two reports by C. R. Dyer and M. Shneier.

ISBN 0-12-267180-5

subset of a 2^n-by-2^n array, which we regard as being made up of unit-square "pixels". Any boundary of such a region can thus be specified, relative to a given starting point, as a sequence of unit vectors in the principal directions. We can represent the directions by numbers, e.g. let i represent 90i° (i=0,1,2,3). For example, the direction sequence for the boundary of the region in Figure 1a, moving clockwise starting from the leftmost of the uppermost border points, is

$$0\ 3\ 0^2\ 3^5\ 2^3\ 1\ 2\ 3^3\ 0\ 3\ 2^5\ 1^6\ 0\ 1\ 0\ 1\ 0\ 3\ 0\ 1\ 0\ 1.$$

This type of boundary representation is called a <u>chain code</u>. Generalized chain codes, involving more than four directions, can also be used. Chain codes provide a very compact region representation, and make it easy to detect features of the region boundary, such as sharp turns ("corners") or concavities. On the other hand, it is harder to determine properties such as elongatedness from a chain code, and it is also difficult to perform operations such as union and intersection on regions represented by chain codes. A general introduction to chain codes and their uses can be found in (1).

Another class of region representations involves various types of maximal "blocks" that are contained in a given region. For example, we can represent a region R as a linked list of the runs (of pixels) in which R meets the successive rows of the array (2). Here each "block" is a 1-by-m rectangle, where m is the run length; the runs are the largest such blocks that R contains, and R is determined by specifying the initial points (or centers) and lengths of the runs. Alternatively, we can represent R by the set of maximal square blocks (or blocks of any other desired shape) that it contains; here R is determined by specifying the centers and

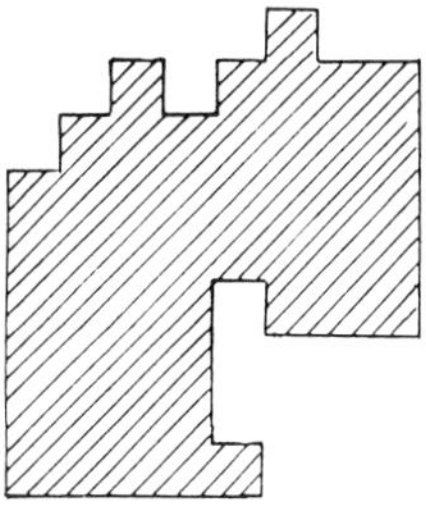

a. Region.

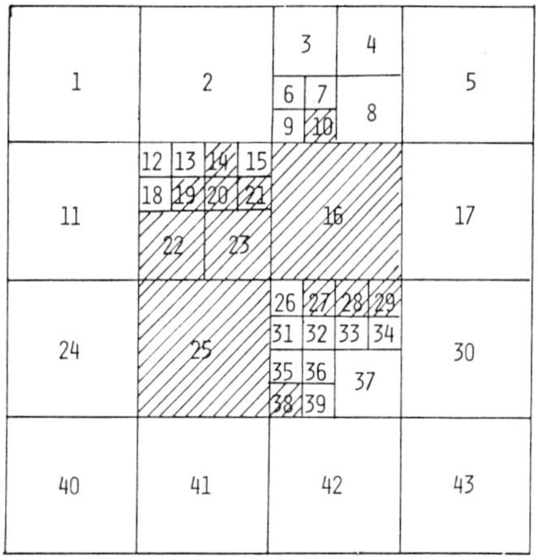

b. Block decomposition of the region in (a).

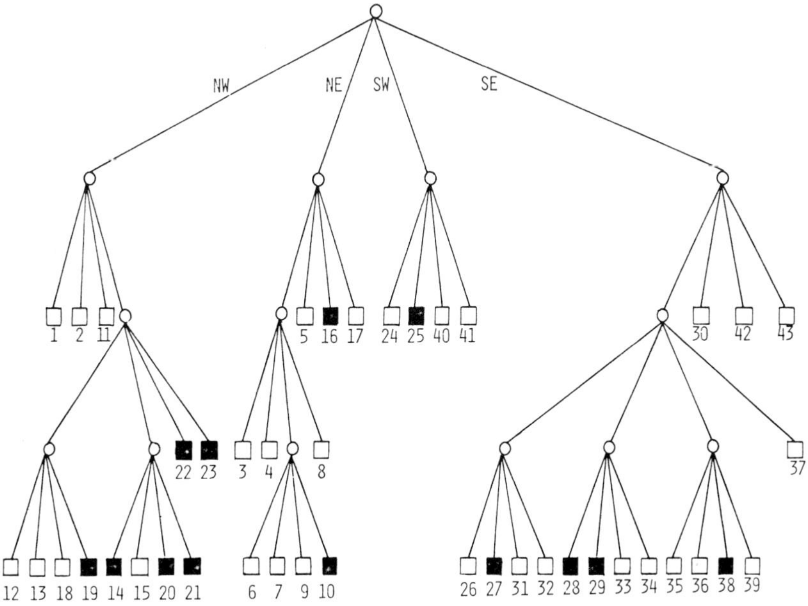

c. Quadtree representation of the blocks in (b).

FIGURE 1. A region, its maximal blocks, and the corresponding quadtree. Blocks in the region are shaded, background blocks are blank.

radii of these blocks. This representation is called the medial axis transformation, or MAT (3). It is somewhat less compact than chain code (4), but it has advantages with respect to performing union and intersection operations or detecting properties such as elongatedness (in terms of the smallness of the radii relative to the number of centers).

There has been recent interest in an approach to region representation based on successive subdivision of the array into quadrants. If the region does not cover the entire array, we subdivide the array, and repeat this process for each quadrant, each subquadrant,... as long as necessary, until we obtain blocks (possibly single pixels) that are entirely contained in the region or entirely disjoint from it. The resulting blocks for the region of Figure 1a are shown in Figure 1b. This process can be represented by a tree of degree 4 (for brevity: a quadtree) in which the entire array is the root node, the four sons of a node are its quadrants, and the leaf nodes correspond to those blocks for which no further subdivision is necessary.[1] The quadtree representation for Figure 1b is shown in Figure 1c. Note that here again we are representing the region as a union of maximal blocks, but this time the blocks must have standard sizes and positions (powers of 2). Since the array was assumed to by 2^n-by-2^n, the tree height is at most n. This method of region representation was proposed by Klinger (6-8); it has also been

[1]The quadtree region representation described here should not be confused with the quadtree representation of two-dimensional point data introduced by Finkel and Bentley (5).

studied extensively by Hunter and Steiglitz (9-11). It is relatively compact, and is also well suited to operations such as union and intersection, and to detecting various region properties.

This paper informally describes a collection of algorithms for converting between quadtree and other representations, and for measuring geometric properties of regions represented by quadtrees. Detailed descriptions of the algorithms can be found in a series of papers by Samet et al. (12-20).

2. CONVERSION

2.1. Quadtrees and Arrays

In Section 1 we described a method of constructing the quadtree corresponding to a given region by recursively subdividing the picture into blocks which are quadrants, subquadrants,... . We assume that the region is represented by a binary array, with region points having value 1 and nonregion points value 0. If a block consists entirely of 1's or 0's, it corresponds to a "gray" nonleaf node, which has four sons corresponding to its four quadrants. If we do this in a "top-down" fashion, i.e., first examine the entire picture, then its quadrants, then their quadrants, etc. as needed, it may require excessive computational effort, since parts of the picture that contain finely divided mixtures of 0's and 1's will be examined repeatedly.

As an alternative, we can build the quadtree "bottom up" by scanning the picture in a suitable order, e.g. in the sequence

1	2	5	6	17	18	21	22
3	4	7	8	19	20	23	24
9	10	13	14	25	26	29	30
11	12	15	16	27	28	31	32
33...							

where the numbers indicate the order in which the points are examined. As we discover maximal blocks of 0's or of 1's, we add leaf nodes to the tree, together with their needed ancestor gray nodes. This can be done in such a way that leaf nodes are never created until they are known to be maximal, so that it is never necessary to merge four leaves of the same color and change their common parent node from gray to black or white. For the details of this algorithm see (17).

Bottom-up quadtree construction becomes somewhat more complicated if we want to scan the picture row by row. Here we add leaf nodes to the tree as we discover maximal 1-by-1 or 2-by-2 blocks of 0's or 1's; if four leaves with a common father all have the same color, they are merged. The details can be found in (16).

Given a quadtree, we can construct the corresponding binary picture by scanning the tree and, for each leaf, creating a block of 0's or 1's of the appropriate size in the appropriate position. A more complicated process can be used if we want to create the picture row by row. Here we must visit each quadtree node once for each row that intersects it (i.e., a node corresponding to a 2^k by 2^k block is visited 2^k times), and, for each leaf, output a run of 0's or 1's of the appropriate length (2^k) in the appropriate position. For the details, see (18).

2.2. Quadtrees and Borders

In order to determine, for a given leaf node M of a quadtree, whether the corresponding block is on a border, we must visit the leaf nodes that correspond to 4-adjacent blocks and check whether they are black or white. To find the nodes corresponding to, e.g., right-hand neighbor blocks, we move upward from M in the tree until we reach some ancestor node from its northwest or southwest son. (If we reach the root node before this happens, M is on the east edge of the picture and has no right-hand neighbor blocks.) As soon as this occurs, we go back down the tree making the mirror images of the moves made on the way up--i.e., the first move down is to the northeast or southeast son, and the following moves are to the northwest or southwest sons. If a leaf node is reached by the time we come to the end of this move sequence, its block is at least as large as M's block, and so is M's sole right-hand neighbor. Otherwise, the non-leaf node reached at the end of the sequence is the root of a subtree whose leftmost leaf nodes correspond to M's right-hand neighbors, and we can find these nodes by traversing that subtree.

Let M,N be black and white leaf nodes whose blocks are 4-adjacent. Thus the pair M,N defines a common border segment of length 2^k (the smaller of the side lengths of M and N) which ends at a corner of M or of N (or both). To determine the next segment along this border, we must find the other leaf P whose block touches the end of this segment:

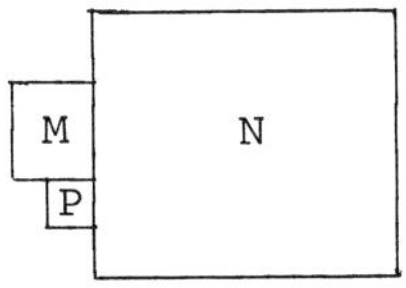

If the segment ends at a corner of both M and N, we must find the other two leaves P,Q whose blocks meet at that corner:

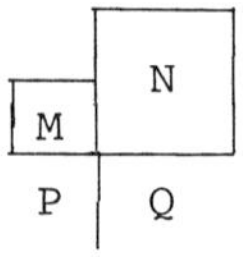

This can be done by an ascending and descending procedure similar to that described in the preceding paragraph; see (12) for the details. The next border segment is then the common border defined by M and P if P is white, or by N and P if P is black. (In the common corner case, the pair of blocks defining the next border segment is determined exactly as in the standard "crack following" algorithm for traversing region borders.) This process is repeated until we come to M,N again, at which stage the entire border has been traversed. The successive border segments constitute a 4-direction chain code, broken up into pieces whose lengths are powers of 2. The time required for this process is on the order of the number of border nodes times the tree height.

Using the methods described in the last two paragraphs, we can traverse the quadtree, find all borders, and generate their codes. We should mark each border as we follow it, so that we will not follow it again from another starting point; note that the marking process is complicated by the fact that a node's block may be on many different borders.

To generate a quadtree from a set of 4-direction chain codes, we first traverse each code and create pairs of leaf nodes having the given border segments, together with the necessary non-leaf nodes. We then generate leaf (and non-leaf) nodes corresponding to the interior blocks. At any stage, if four leaves with a common father all have the same color, they are merged. The details of this algorithm will not be given here; see (13). The time required is on the order of the perimeter (=total 4-direction chain code length) times tree height.

2.3. Quadtrees of Derived Sets

Let S be the set of 1's in a given binary array, and let $\overline{S}$ be the complement of S. The quadtree of $\overline{S}$ is the same as that of S, with black leaf nodes changed to white and vice versa. To get the quadtree of $S \cup T$ from those of S and T, we traverse the two trees simultaneously. Where they agree, the new tree is the same. If S has a gray (=nonleaf) node where T has a black node, the new tree gets a black node; if T has a white node there, we copy the subtree of S at that gray node into the new tree; if S has a white node and T a black node, the new tree gets a black node. The algorithm for $S \cap T$ is exactly analogous, with the roles of black and white reversed. The time required for these algorithms is proportional to the number of nodes in the smaller of the two trees (20).

3. PROPERTY MEASUREMENT

3.1. Connected Component Labeling

Given a binary array represented by a quadtree, we can label its components by traversing the tree in a standard order, say NW, NE, SW, SE. Whenever we come to a black leaf node, we visit the leaf nodes whose blocks adjoin M's block on its south and east sides (and at its southeast corner, if we are labeling 8-components); see Section 2.2 on how to find these nodes. If we find unlabeled black leaf nodes, we give them the same label as M; if we find black leaf nodes that already have labels, we note that their labels are equivalent. When the traversal is complete, we sort out the equivalences, retraverse the tree, and give the black leaf nodes their final labels. The time required is on the order of the number of nodes in the tree times the tree height. Alternatively, it is on the order of Q log Q, where Q is the number of leaf nodes in the tree. For the details of this algorithm see (15).

3.2. Component Counting and Genus Computation

Once the connected components have been labeled, it is trivial to count them, since their number is the same as the number of inequivalent labels. We will next describe a method of determining the number of components minus the number of holes by counting certain types of local patterns in the array; this number g is known as the genus or Euler number of the array.

Let V be the number of 1's, E the number of 11's and $\begin{smallmatrix}1\\1\end{smallmatrix}$'s, and F the number of $\begin{smallmatrix}11\\11\end{smallmatrix}$'s in the array; it is well known (21) that the $g = V - E + F$. This result can be generalized to the case where the array is represented by a quadtree (19). In fact, let V be the number of black leaf nodes; E the number of pairs of such nodes whose blocks are horizontally or vertically adjacent; and F the number of triples or quadruples of such nodes whose blocks meet at and surround a common point, e.g.

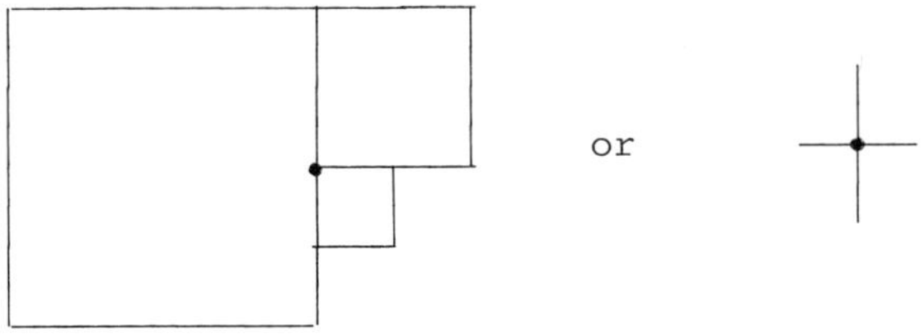

Then g = V - E + F. These adjacencies can be found (see Section 2.2) by traversing the tree; the time required is on the order of the number of nodes in the tree times the tree height, or of Q log Q, as before.

3.3. Area and Moments

The area of a region represented by a quadtree can be obtained by summing the areas of the black leaf nodes, i.e., counting 4^h for each such node that represents a 2^h by 2^h block. Similarly, the first x and y moments of the region relative to a given origin can be computed by summing the first moments of these blocks; note that we know the position (and size) of each block from the position of its leaf in the tree. Knowing the area and the first moments gives us the coordinates of the centroid, and we can then compute central moments relative to the centroid as origin. The time required for any of these computations is essentially proportional to the number of nodes in the tree. Further details on moment computation from quadtrees can be found in (20).

3.4. Perimeter

The perimeter of a region represented by a quadtree can be obtained by traversing all the borders of the region (Section 2.2) and summing the numbers of steps in the resulting codes. Alternatively, it can be computed by traversing the tree, and for each leaf node, checking the colors of the nodes whose blocks are adjacent to its block on two sides, say bottom and right, to determine which of these adjacencies contributes to the total perimeter. The time required for this is proportional to the number of leaf nodes; see (14) for the details.

4. CONCLUDING REMARKS

Quadtrees constitute an interesting alternative to the standard methods of digitally representing regions. Their chief disadvantage is that they are not shift-invariant; two regions differing only by a translation may have quite different quadtrees. Thus shape matching from quadtrees is not straightforward. In other respects, however, they have many potential advantages. They provide a compact and easily constructed representation from which standard region properties can be efficiently computed. In effect, they are "variable-resolution arrays" in which detail is represented only when it is available, without leading to excessive storage requirements for parts where detail is lacking. Hopefully, this paper has drawn attention to the advantages of quadtrees as data structures for possible use in digital cartographic information systems.

REFERENCES

1. H. Freeman, Computer processing of line-drawing images, Computing Surveys 6, 1974, 57-97.
2. D. Rutovitz, Data structures for operations on digital images, in G. C. Cheng et al., eds., Pictorial Pattern

Recognition, Thompson Book Co., Washington, DC, 1968, 105-133.

3. H. Blum, A transformation for extracting new descriptors of shape, in W. Wathen-Dunn, ed., Models for the Perception of Speech and Visual Form, M.I.T. Press, Cambridge, MA, 1967, 362-380.

4. J. L. Pfaltz and A. Rosenfeld, Computer representation of planar regions by their skeletons, Comm. ACM 10, 1967, 119-122, 125.

5. R. A. Finkel and J. L. Bentley, Quadtrees: a data structure for retrieval on composite keys, Acta Informatica 4, 1974, 1-9.

6. A. Klinger, Data structures and pattern recognition, Proc. 1st Intl. Joint Conf. on Pattern Recognition, 1973, 497-498.

7. A. Klinger and C. R. Dyer, Experiments in picture representation using regular decomposition, Computer Graphics Image Processing 5, 1976, 68-105.

8. N. Alexandris and A. Klinger, Picture decomposition, tree data-structures, and identifying directional symmetries as node combinations, ibid. 8, 1978, 43-77.

9. G. M. Hunter, Efficient computation and data structures for graphics, Ph.D. dissertation, Dept. of Electrical Engineering and Computer Science, Princeton University, Princeton, NJ, 1978.

10. G. M. Hunter and K. Steiglitz, Operations on images using quad trees, IEEETPAMI-1, 1979, 145-153.

11. G. M. Hunter and K. Steiglitz, Linear transformations of pictures represented by quad trees, Computer Graphics Image Processing 10, 1979, 289-296.

12. C. R. Dyer, A. Rosenfeld, and H. Samet, Region representation: boundary codes from quadtrees, Computer Science Center TR-732, University of Maryland, College Park, MD, February 1979.

13. H. Samet, Region representation: quadtrees from boundary codes, Computer Science Center TR-741, University of Maryland, College Park, MD, March 1979.

14. H. Samet, Computing perimeters of images represented by quadtrees, Computer Science Center TR-755, University of Maryland, College Park, MD, April 1979.

15. H. Samet, Connected component labeling using quadtrees, Computer Science Center TR-756, University of Maryland, College Park, MD, April 1979.

16. H. Samet, Region representation: raster-to-quadtree conversion, Computer Science Center TR-766, University of Maryland, College Park, MD, May 1979.

17. H. Samet, Region representation: quadtrees from binary arrays, Computer Science Center TR-767, University of Maryland, College Park, MD, May 1979.

18. H. Samet, Region representation: quadtree-to-raster conversion, Computer Science Center TR-768, University of Maryland, College Park, MD, June 1979.

19. C. R. Dyer, Computing the Euler number of an image from its quadtree, Computer Science Center TR-769, University of Maryland, College Park, MD, June 1979.

20. M. Shneier, Linear time calculations of geometric properties using quadtrees, Computer Science Center TR-770, University of Maryland, College Park, MD, May 1979.

21. A. Rosenfeld and A. C. Kak, Digital Picture Processing, Academic Press, New York, 1976.

ANALYSIS AND MANIPULATION OF LINEAL MAP DATA

Herbert Freeman

Rensselaer Polytechnic Institute
Troy, New York USA

Algorithms are described for analyzing and manipulating 8-point chain-encoded lineal map data. The convenience of the chain representation for yielding to efficient algorithms is demonstrated.

1. INTRODUCTION

The form of representation selected for line-drawing data can have a significant effect on the relative ease with which the encoded data can be processed in a digital computer. Ease of processing is, of course, not the only consideration in selecting a coding scheme - precision, ease of encoding, storage efficiency, and smoothness of display are other factors of importance. However, where extensive processing is required, the facility with which it can be accomplished is often an overriding concern.

The familiar 8-direction, square grid-based chain code (1, 2) illustrated in Figure 1 has been found to lend itself well to efficient line-drawing processing, especially for applications involving the form or shape of the encoded contours. (The algorithms are less effective, though still workable, for

ISBN 0-12-267180-5

Figure 1. A chain-encoded contour map.

problems that are area-oriented, that is, problems involving the area enclosed by the encoded contour rather than the contour itself.) A large variety of algorithms has been developed for processing irregular curves encoded with the 8-point chain code. For a detailed listing and description of these the reader is referred to the literature (2). However, a few of the algorithms will be briefly summarized here, together with some new ones not previously published.

2. ALGORITHMS

Let us consider a simple 8-point chain A. It is represented by

$$A = a_1 a_2 \dots a_n, \qquad a_i \in \{0, \dots, 7\} \tag{1}$$

where each link a_i is a short vector of length 1 (i = even) or length $\sqrt{2}$ (i = odd) for a unity-spaced uniform grid. Each link a_i can be resolved into components along one of the 4 permissible directions $m \cdot 45^\circ$ (m = 0,..,3). Such components are denoted by a_{im}.

Information for linking individual chains, the insertion of markers and the inclusion of other "house-keeping" data is handled by means of special signal codes. They are identified by the prefix 04, followed by two or more octal digits. Thus 0400 is the end-of-chain indicator, 0401 indicates the beginning of an "invisible" connecting chain (inserted solely to maintain the relative positioning of two chains), and 0402 indicates the end of such an "invisible" connecting chain. A valid 04 link sequence is denoted by 0404. A marker can be indicated by 0405XYZ where XYZ is the 3-digit octal marker number. An extensive listing of these signal codes can be found elsewhere (2).

A. Inverse of chain

$$(a_1 \dots a_n)^{-1} = a_n^{-1} \dots a_1^{-1} \tag{2}$$

where

$$a_i^{-1} = a_i \dot{+} 4 \tag{3}$$

and the symbol $\dot{+}$ indicates modulo-8 addition. The inverse of a chain describes the same curve as the chain but in the oppositely directed sense.

B. Length of a Chain

$$L = n_e + n_o \sqrt{2} \tag{4}$$

where n_e is the number of even-valued and n_o is the number of odd-valued links in the chain.

For large chains, the length is approximately equal to 1.17 times the total number of links in the chain. This approximation is based on the assumption that for large chains about 41 percent of the links are odd-valued (3,4).

C. Span E^m (width in direction m, m = 0,1,2,3)

$$E^m = \max_j w_j^m - \min_j w_j^m, \quad (j = 0,1,\ldots,n) \tag{5}$$

where

$$w_j^m = \sum_{i=1}^{j} a_{im} \tag{6}$$

$$w_o^m = 0 \qquad \text{for all } m \tag{7}$$

and the a_{im} are the projections of the vector a_i on the m-axes, m = 0,1,2,3. They are given in the following table:

a_i	a_{i0}	a_{i1}	a_{i2}	a_{i3}
0	1	$\sqrt{2}/2$	0	$-\sqrt{2}/2$
1	1	$\sqrt{2}$	1	0
2	0	$\sqrt{2}/2$	1	$\sqrt{2}/2$
3	-1	0	1	$\sqrt{2}$
4	-1	$-\sqrt{2}/2$	0	$\sqrt{2}/2$
5	-1	$-\sqrt{2}$	-1	0
6	0	$-\sqrt{2}/2$	-1	$-\sqrt{2}/2$
7	1	0	-1	$-\sqrt{2}$

D. Integration with Respect to x Axis (m=0)

$$S = \sum_{i=1}^{n} a_{i0} (y_{i-1} + \frac{1}{2} a_{i2}) \tag{8}$$

where

$$y_i = y_{i-1} + a_{i2} \tag{9}$$

and y_0 is the ordinate of the initium (starting node of chain).

For a closed chain (initium = terminus), y_0 can be arbitrarily selected and the formula will yield the net area encircled in a clockwise sense (interior toward right).

Observe that since a_{i0} and a_{i2} can take on values of only 1, 0, and -1, the determination of area can be programmed using only integer additions and subtractions; multiplication is not required.

E. First Moment About x Axis (m = 0)

$$M_1^x = \sum_{i=1}^{n} \frac{1}{2} a_{i0}[y_{i-1}^2 + a_{i2}(y_{i-1} + \frac{1}{3} a_{i2})] \tag{10}$$

Here, as in the formula for eq. (8), it is readily possible to formulate the algorithm so that only integer additions, integer subtractions, and table-lookup operations are required, (with the exception of some initial and final multiplications and divisions for initialization and final rescaling). The quadratic term in y_{i-1} is computed using

$$y_i^2 = y_{i-1}^2 + 2y_{i-1} a_{i2} + a_{i2}^2 \tag{10a}$$

$$y_i = y_{i-1} + a_{i2} \tag{10b}$$

and the divisions by 2 and 3 at each step are avoided by multiplying the entire expression by 6. A similar technique can be applied in the case of the second moment, for which the expression is given next, as well as for any higher-order moments.

F. Second Moment About x Axis (m = 0)

$$M_2^x = \sum_{i=1}^{n} \frac{1}{3} a_{i0}[y_{i-1}^3 + \frac{3}{2} a_{i2} y_{i-1}^2 + a_{i2}^2 y_{i-1} + \frac{1}{4} a_{iy}] \quad (11)$$

G. Centroid

The centroid of a closed chain is readily computed using the area and first-moment algorithms:

$$x_c = -M_1^y/S$$
$$y_c = M_1^x/S \quad (12)$$

where M_1^y, the first moment about the y axis, is computed by first rotating the chain by $\pi/2$ about the origin and then using eq. (10). Rotation by $\pi/2$ is obtained by rotating the coordinates of the initium and then adding 2 modulo 8 to each link of the chain. (A change in sign is also required.)

H. Center of Gyration

The computation for the center of gyration is analogous to that for the centroid except that the second moment M_2^x is used.

The first, second and possibly higher-order moments provide useful features for classification of line-drawing patterns. The centroid (12) can serve as a robust, noise-insensitive reference point for a contour.

I. Distance Between Two Points

$$d = [(\sum_{i=1}^{n} a_{i0})^2 + (\sum_{i=1}^{n} a_{i2})^2]^{1/2} \quad (13)$$

where $\{a_i\}_1^n$ is any chain connecting the two points.

J. Residue Chain

The residue chain is the shortest chain connecting two points, or, relative to another chain, the shortest chain that connects the initium to the terminus of that chain. It can be determined by a procedure of equivalent-link substitutions and cancellations (2). An alternative procedure is as follows:

$$w_n^0 = \sum_{i=1}^{n} a_{i0}$$
$$w_n^2 = \sum_{i=1}^{n} a_{i2} \qquad (14)$$

$$n_e = |w_n^0| - |w_n^2| \qquad (15)$$

$$n_R = \max.(|w_n^0|, |w_n^2|) \qquad (16)$$

$$n_0 = n_R - n_e \qquad (17)$$

where n_e and n_0 are, respectively, the number of even-valued and odd-valued links in the n_R-link residue chain.

The residue chain itself will consist of the string of n_e links of value L_e concatenated with n_0 links of value L_0, where L_e and L_0 are given by the following table:

w_n^0	w_n^2	$w_n^0 - w_n^2$	Octant	L_e	L_0
+	+	+	1	0	1
+	+	-	2	2	1
+	-	+	8	0	7
+	-	-	7	6	7
-	+	+	4	4	3
-	+	-	3	2	3
-	-	+	5	4	5
-	-	-	6	6	5

(The + and - symbols indicate the signs of the terms in the column heading.)

K. Distance From a Line

Suppose we are given a chain $\{a_i\}_1^n$ and wish to find the minimum distance from this chain to a specified line $y = px+q$. The problem is illustrated in Figure 2.

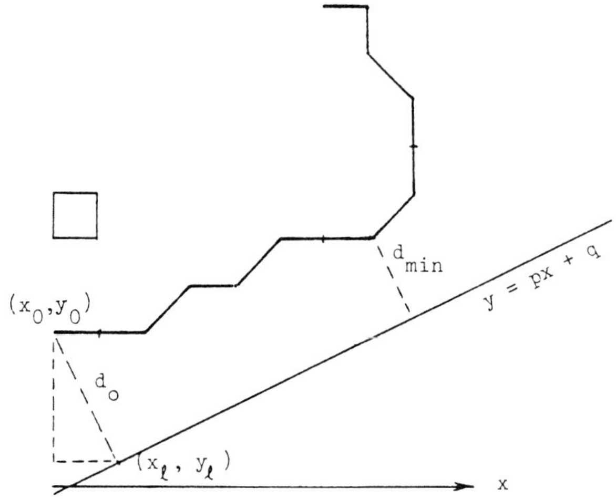

Figure 2 Determination of Minimum Distance Between a Chain and a Line

We first find the distance from the chain initium (x_0, y_0) to the line. The equation of the perpendicular line passing through the point (x_0, y_0) is

$$y = \frac{1}{p}(x_0 - x) + y_0 \tag{18}$$

Simultaneous solution of the two line equations yields the intersection (x_ℓ, y_ℓ):

$$x_\ell = \frac{p}{1+p^2}\left(y_0 + \frac{1}{p}x_0 - q\right) \tag{19}$$

$$y_\ell = \frac{p^2}{1+p^2}\left(y_0 + \frac{1}{p}x_0 - q\right) + q \tag{20}$$

and the initial distance

$$d_0 = [(x_0 - x_\ell)^2 + (y_0 - y_\ell)^2]^{1/2} \tag{21}$$

Next we find the projections a_{is} of the chain links a_i on the line perpendicular to y = px + q. They are as follows (see Figure 3):

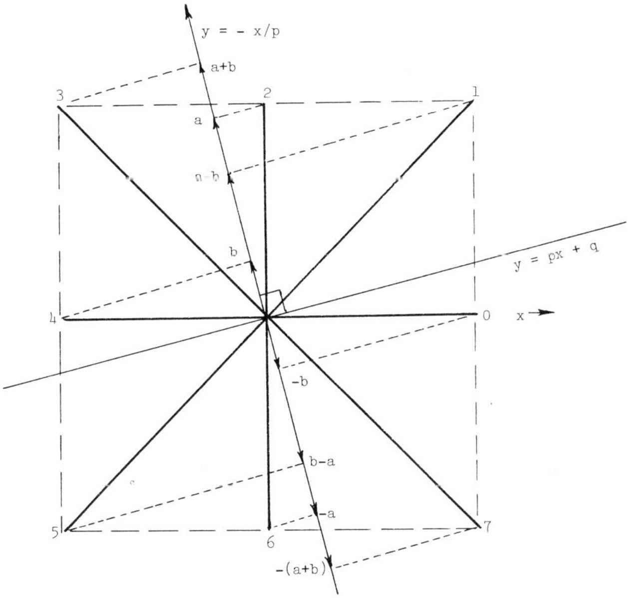

Figure 3. Computation of Chain Link Projections

a_i	a_{is}	a_i	a_{is}
0	-b	4	b
1	a-b	5	b-a
2	a	6	-a
3	a+b	7	-(a+b)

where

$$a = 1/\sqrt{1+p^2} \tag{22}$$

$$b = p/\sqrt{1+p^2} \tag{23}$$

The distance d_i from the line to the end node of link a_i is given by

$$d_i = d_0 + \sum_{s=0}^{i} a_{is} \quad (i = 1,\ldots,n) \tag{24}$$

and the desired minimum distance is simply

$$\min_i . \{d_i\}$$

L. Intersections

Suppose we are given two chains and wish to determine their intersections, if any. The best algorithm for this purpose is the one suggested by Morse (5). We determine the bounding x-y rectangle for each chain (formed by x_{min}, x_{max}, y_{min}, and y_{max}). Any intersections of the two chains must lie in a rectangular area common to the two bounding rectangles. If no such area exists, the two chains do not intersect.

If there is a common rectangular area, one discards the portions of the two chains lying outside this area, and, for the remaining chain portions, again finds bounding rectangles for each, after which the entire process is repeated. If, as is likely, a chain separates into multiple, disjoint portions, separate rectangles are found for each until either it is established that there is no intersection or the intersection(s) are found. The process is illustrated in Figure 4a, b, c, and d.

The above algorithm may come to a "dead end" if at some point the two bounding rectangles overlap perfectly (see Figure 5a). To overcome this difficulty one divides the common rectangular area in half and continues the process for the two

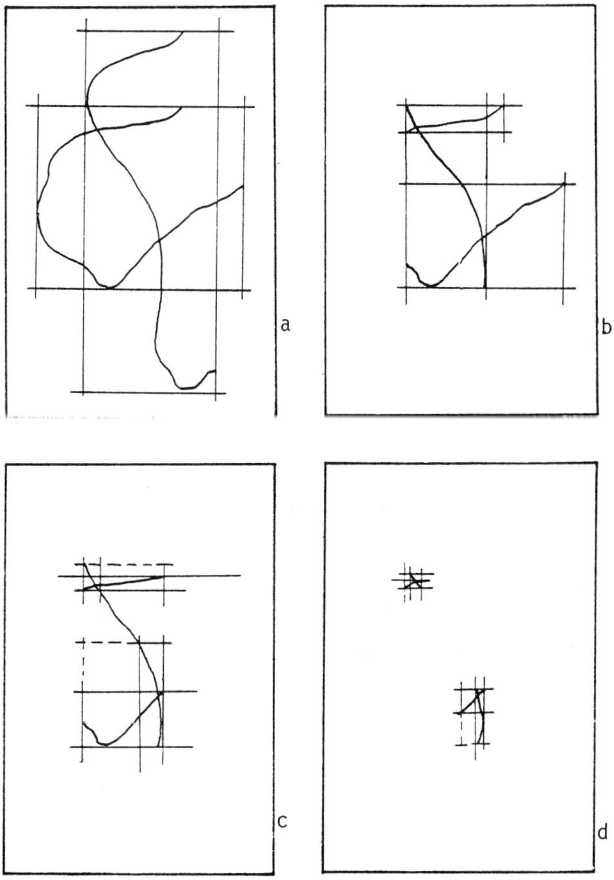

Figure 4. Illustration of Four Iterations of Intersection Algorithm.

halves separately. Alternate vertical and horizontal halving is recommended. This scheme is illustrated in Figure 5b.

M. Encirclement of a Point

The following algorithm enables us to determine whether a point lies inside, on, or outside a closed chain-coded contour. Given a chain, together with the x-y coordinates of its initium and a specified point (x_o, y_o), we draw a horizontal line from

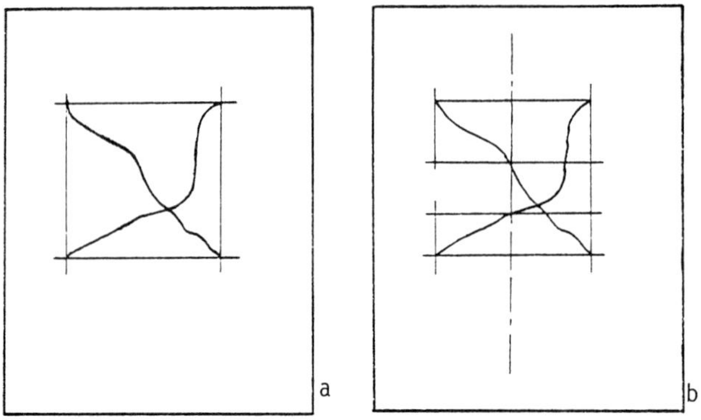

Figure 5. Resolving Intersection Algorithm "Lock-Up" by Halving.

$(x_o, y_0 + 1/2)$ to the right boundary of the picture. If the point (x_o, y_o) lies inside the closed contour, the line from $(x_o, y_o + 1/2)$ must intersect the contour an odd number of times. (The horizontal line is drawn from $(x_o, y_o + 1/2)$ rather than from (x_o, y_o) to prevent the possibility of having chain links lie on the line.)

We now trace out the chain once completely, node by node, starting from the initium. Every time the y-coordinate changes from $y_o + 1/2$ to $y_o - 1/2$ or vice versa we check the x-coordinate. If the x-coordinate is greater than x_o, we increment a counter. When the chain has been completely traced, if the count is odd, the point lies inside the chain; if it is even, it lies outside. If one of the nodes had a value equal to x_o, the point lies on the chain. The algorithm is illustrated in Figure 6.

It is important to note that with this algorithm the entire chain must be traced out. If the contour representation is first converted from the chain code to Merrill's Y-partition

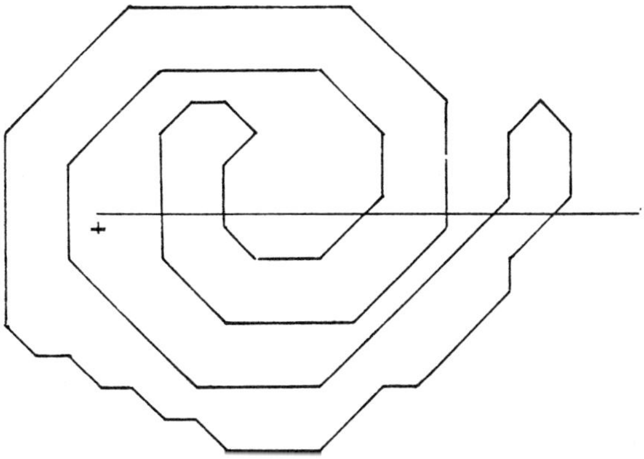

Figure 6. Determining Whether Point Lies Inside, On, or Outside Chain Contour.

code (6), the computation is much faster since only the odd-even pairs in the y-partition for y_o must be checked. This points out the well-known fact that the chain representation is shape-oriented; that is, it is well-suited for problems involving the shape of the described contour. For area-oriented problems, Merrill's Y-partition code, though much less efficient in storage, is more suitable and yields faster algorithms. For such problems conversion from the chain code to the Y-partition code may be a worthwhile price to pay.

N. Perpendicular to a Chain at a Point

To determine the perpendicular to a chain at the terminus of some chain link a_u, we average over an even number of links, say 2k, to smooth over the quantization noise. The value of k will normally range from 2 to 5 for a well-quantized curve. Let

$$D_x(u) = \sum_{j=-k}^{k+1} a_{u+j,0}$$

$$D_y(u) = \sum_{j=-k}^{k+1} a_{u+j,2} \tag{25}$$

The angle the perpendicular makes with the horizontal at the terminus of a_u will be

$$\phi(u) = -\tan^{-1}[D_x(u)/D_y(u)] \tag{26}$$

O. Generation of Perpendicular Chain (Gradient Chain)

We can also generate a chain that assumes a form essentially orthogonal to that of a given chain at a specified node. Such an "orthogonal" chain is useful for representing the gradient in a contour map.

We shall follow the convention of "interior or higher elevation to the right" in defining the directions for all contour chains. The gradient chain will accordingly be directed inward (i.e., to the right).

Given a chain link a_i, its inward perpendicular link a_i^p will be given by a $a_i \mathbin{\dot{-}} 2$. The perpendicular (or gradient) chain at the terminus of a_i can now be defined as

$$a_i^p \; a_{i+1}^p \; a_{i-1}^p \; a_{i+2}^p \; a_{i-2}^p \; a_{i+3}^p \; \cdots$$

Figure 6 shows a set of three contour lines and a computed gradient line. As used here the gradient chain is "orthogonal" only to the contour chain from which it departs to the right, not to the contour chain which it approaches from the left. If this approximation is objectionable, the gradient can be improved by also computing a left perpendicular chain back from the terminating node and replacing the second half of the links of the gradient chain with links from this left gradient chain, making due adjustments to assure that the gradient will still connect one contour to the other. An illustration of such an improved "right-left" gradient is also shown in Figure 7. If desired, the process can be iterated; but such refinement is seldom necessary.

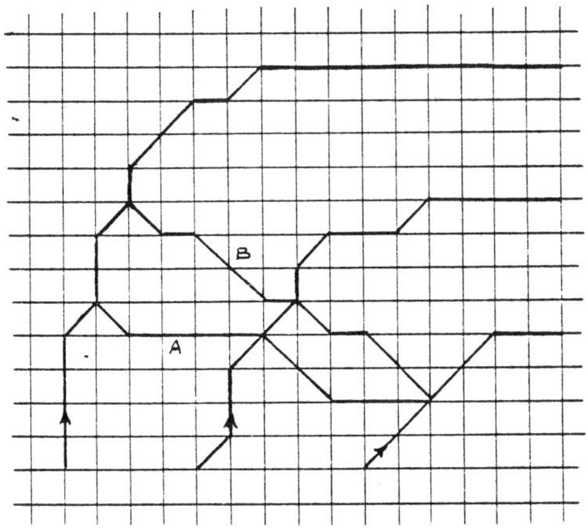

Figure 7. Generation of Gradient Chain (A - Right Perpendicular Only; B - Combined Right and Left Perpendicular).

P. Rotation

One of the most difficult and time-consuming chain processing tasks is that of rotation. Rotation by a multiple of 90 degrees is simple enough; one merely multiplies each link by the appropriate multiple of 2, modulo 8. For other rotations, the chain must be treated as if it were a curve, rotated, and then re-quantized and re-encoded into the underlying lattice. Unfortunately, the process of re-quantization introduces additional quantization noise and thus reduces the precision of the resulting representation. Some of the noise can, however, be removed through judicious post-editing.

Whenever a well-quantized smooth curve is encoded into the chain representation, successive links will normally be either identical or differ by +1 or -1 (modulo 8). Only in rare cases - at corners - will the link-to-link difference be greater in magnitude. However, when such a chain is re-quantized because of rotation, the accompanying noise may manifest itself in the presence of link-to-link differences of +2 or -2 at places in the chain where no "corner" was previously evident. Such noise can be removed by tracing out both the original chain and the rotated chain simultaneously. If a +2 change is found in the rotated chain for which there is no corresponding such change in the original chain, the offending two-link pair is replaced by its single-link equivalent (e.g., a 06 link pair would be replaced by the link 7).

The problem is illustrated in Figure 8. In Figure 8a we show a straight line 1 represented by a chain labelled 2. If the line 1 is rotated 30 degrees, line 3 is obtained, which can be represented by the chain labelled 4.

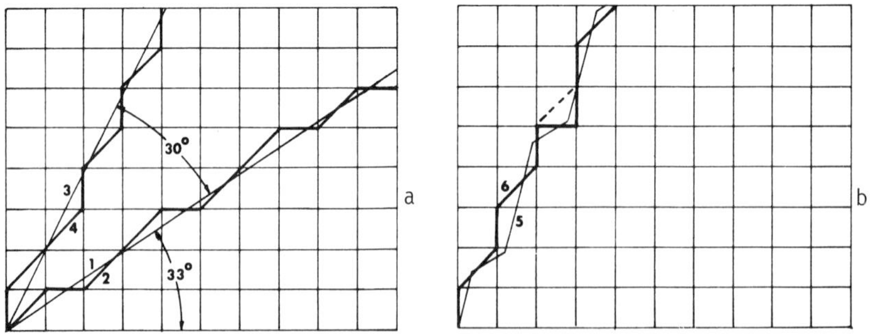

Figure 8. Illustration of Quantization Noise in Chain Rotation Algorithm.

If the chain 2 (rather than line 1) is rotated by 30 degrees, it will appear as curve 5 in Figure 8b. Re-quantization then yields the chain labelled 6. We observe that the chain 6 contains two successive "90-degree turns", representing link differences of -2 and +2 which, of course, were present neither in the original chain 2 nor in the chain of the rotated line, chain 4. The problem is corrected by replacing the link pair 02 by the single link 1, as shown by the dotted line in Figure 8b. The reader will observe that this problem for rotation applies equally to scaling by other than integer multiples and must be handled in the same manner.

3. CONCLUSIONS

Algorithms have been described for processing lineal maps encoded using the familiar 8-point chain coding scheme. They represent only a small subset of all the chain-processing algorithms that have been developed over the years, most of which have been described previously in the literature. Nearly all the algorithms can be programmed for integer arithmetic, using only additions, subtractions, decisions, and table look-up. Processing time tends to be linear with the number of links in the chains. The simplicity with which the algorithms are formulated suggests that algorithms for a large class of other analysis, synthesis, and manipulation tasks can be readily developed as well. The general method is seen to be especially well-suited for free-form line-drawing data such as found in geographic maps and medical imagery, both applications in which it is finding extensive use.

ACKNOWLEDGMENT

This article describes research sponsored by the U.S. Air

Force through the Rome Air Development Center under Contract F30602-78-C-0083. The information presented does not necessarily reflect the position or policy of the U. S. Government and no official endorsement should be inferred.

REFERENCES

1. Freeman, H., "On the Encoding of Arbitrary Geometric Configurations", IRE Trans., EC-10, (2), June 1961, pp. 260-268.
2. Freeman, H., "Computer Processing of Line-Drawing Images", Computing Surveys, 6, (1), March 1974, pp. 57-97.
3. Freeman, H., "A Technique for the Classification and Recognition of Geometric Patterns", Proc. 3rd International Congress on Cybernetics, Namur, Belgium, 1961, pp. 348-368.
4. Groen, F., and Verbeek, P. W., "Freeman-Code Probabilities of Object Boundary Quantized Contours", Comp. Graph. and Image Proc., 7, (3), June 1978, pp. 391-402.
5. Morse, S. P., "Computer Storage of Contour-Map Data", Proc. 23rd Nat'l. Conf., ACM, 1968, pp. 45-51.
6. Merrill, R. D., "Representation of Contours and Regions for Efficient Computer Search", Comm. ACM, 16, (2), Feb. 1973, pp. 69-82.

REPRESENTATION AND RECOGNITION OF CARTOGRAPHIC DATA

Larry S. Davis[1]

Computer Sciences Department
University of Texas
Austin, Texas

1. INTRODUCTION

Shape representation and recognition has, for the last 15 years, been a major research area in computer image understanding. Shape is the principal cue, for example, in applications as diverse as chromosome analysis, recognition of targets (airplanes, tanks, etc.) in reconnaissance imagery and in character recognition. Figure 1 shows a simplified block diagram of the shape recognition process. An input image is first segmented into meaningful areas (see Rosenfeld and Kak (1976) for a discussion of image segmentation techniques). These segments are then individually analyzed and a representation, or description of each segment, is computed. Each representation is then matched against a previously computed database of analogously formatted models. The model with the highest degree of match is then used to name that segment.

[1]This research was supported in part by funds derived from National Science Foundation Grant ENG-7904037.

ISBN 0-12-267180-5

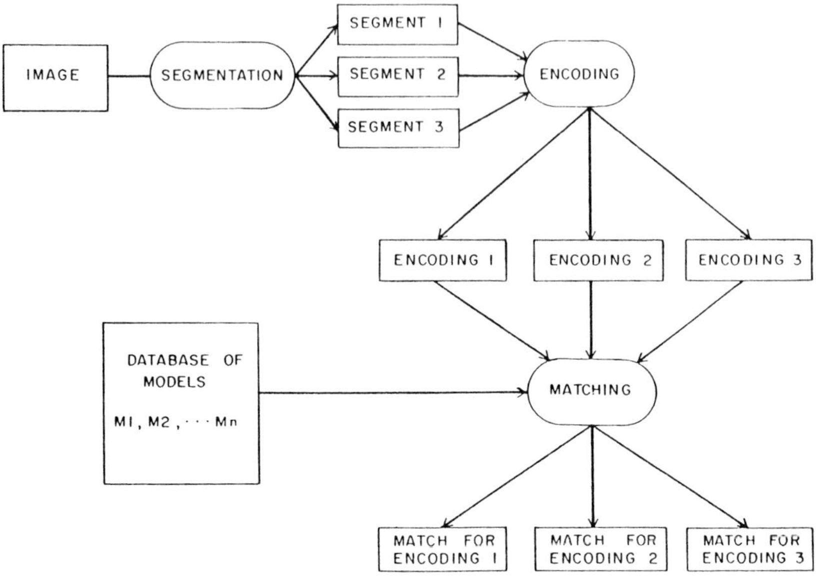

FIGURE 1. Block Diagram of the Shape Recognition Process

Even though a significant amount of energy has been devoted to designing solutions to the shape analysis problem, it is nevertheless true that no general solution exists. Some of the factors to which this can be attributed are:

1) translation: a shape must be recognized independent of its location in the image;

2) orientation: a shape must be recognized independent of its orientation in the image;

3) size, or scale: it is often necessary to recognize a shape over a range of scales;

4) partial information: very often, only a fragment of a shape is available (e.g., a section of a long coastline). It is still necessary to recognize the shape. This problem has been called the segment matching problem (Feder and Freeman 1965);

5) the size of the database: for cartographic applications, the database is likely to be quite large -- several thousand shapes. The database should be organized so that not every shape in the database has to be independently compared against an unknown shape.

Although factors 1-4 have received considerable attention in the computer vision literature, factor 5 has been almost entirely neglected. This can be attributed to the fact that the databases used by computer vision systems are usually quite small, especially when compared to cartographic databases.

This paper will consider two shape representations -- chain codes and polygonal approximations -- and will discuss shape matching algorithms based on each of these. It should be pointed out that factor 4, the need to recognize a shape given only partial information, essentially eliminates from consideration a wide class of shape representations based on global shape descriptors, such as Fourier coefficients (see, e.g., Persoon and Fu 1977) and central moments (see Dudani, et al. 1977). This is because the descriptors for a piece of a shape will not usually bear any simple relationship to the descriptors for the entire shape.

2. CHAIN CODE

The chain code (Freeman 1961) is a slope intrinsic representation of a shape. The chain code of a digitized curve represents the sequences of direction changes between adjacent points on the curve. If the nth point of the curve is at grid position (i,j), then the chain element, or link, corresponding to the change in position from n^{th} point to the $(n+1)^{st}$ point is shown in Figure 2.

A correlation matching procedure for chain coded curves was described by Feder and Freeman (1965). Let $c = c_1c_2 \ldots c_n$ be the chain code of a shape in the database, and let $d = d_1d_2 \ldots d_m$, $m \leq n$, be the chain code of some unidentified curve. Then the <u>chain correlation</u> of d at C_j is:

$$C(j) = \frac{1}{m} \sum_{i=0}^{m} (\cos([d_i - c_{i+j}] \bmod 8 * \frac{\pi}{4})$$

If the chain d matches chain c exactly (link for link) at position j, then $C(j) = 1$. With respect to factors 1-4 discussed in the introduction, we note the following:

1) Chain codes are translation invariant. Translating a digital curve by an integral (with respect to the underlying grid) translation leaves the individual chainlets unchanged. Non-integral translations, which require a redigitization, result in only minor changes in the chain code, which do not seem to affect the matching process.

2) Since the chain code is a slope intrinsic representation, it is not rotation invariant. In fact, rotating a

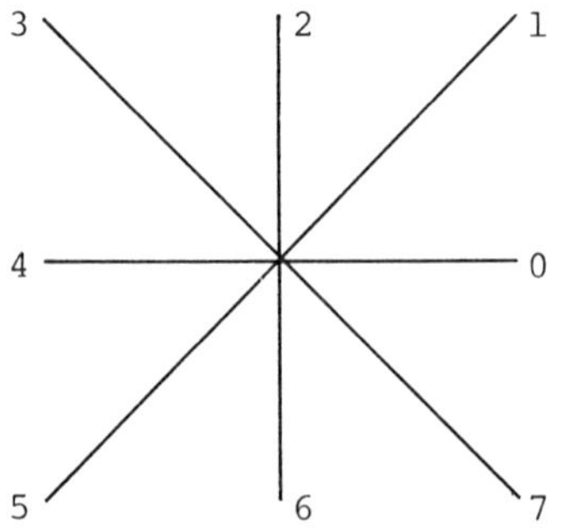

FIGURE 2. Chain Code

curve can even change the length of the chain code. The rotation variance can, to a certain extent, be overcome by computing a curvature intrinsic representation based on chain code differences. Here, if $c_1c_2...c_n$ is the original chain code, then $c'_1c'_2...c'_{n-1}$ is the chain difference code, with $c'_i = |c_{i+1} - c_i| \bmod 8$. The chain difference codes can be compared using a correlation function similar to (1). Feder and Freeman (1965) report that matching chain difference codes is more susceptible to noise than matching chain codes.

3) Chain codes are not size invariant. If precise scale information is available, then a chain code from the database can be suitably averaged and sampled so that its scale is comparable to the scale of the unknown curve. In the absence of such scale information, chain code correlation cannot be applied.

4) Chain codes are, obviously, particularly well suited for matching given only partial information.

Chain code correlation is also computationally quite efficient. Matching a chain of n elements against a chain of m elements requires 0(mn) operations. The 8 values of the cosine function can be conveniently stored in a table. Therefore, in situations where size and orientation can be controlled, chain code correlation offers a robust and inexpensive matching capability.

3. POLYGONAL APPROXIMATIONS

Polygonal approximations provide a representation that will support matching even in the absence of scale and orientation information. Section 3.1 discusses some simple methods for computing polygonal approximations. Section 3.2 presents a matching procedure based on polygonal approximations, and Section 3.3 describes an experiment using cartographic data.

3.1 Computing Polygonal Approximations

Given a chain code representation of a digital curve, computing a polygonal approximation means to segment the curve into pieces which are well fit by straight lines. As in image segmentation, two broad classes of techniques for computing such segmentations can be identified:

1) techniques which directly compute the boundaries between segments. The segments themselves are then the section of curve between two consecutive detected boundaries.

2) techniques which directly compute the segments.

The first class of techniques are angle detection techniques, and are analogous to the edge protection procedures used in image segmentation (see Davis (1975) for a survey). The second class of techniques are functional approximation techniques, and are analogous to the region growing procedures used in image segmentation (see Zucker (1976) for a survey).

A tremendous number of shape segmentation procedures, based on both angle detection and on functional approximation, have been proposed. We will just briefly discuss simple angle detection and functional approximation procedures. A more detailed discussion can be found in Pavlidis (1977).

The angle detection procedure is from Rosenfeld and Kak (1976). Suppose $(x_i, y_i)_{i=0}^{n}$ is a vector representation of a curve, S. At each point in S we compute a number related to the curvature of S at that point:

$$C_k(i) = (\vec{a}_{i_k} \cdot \vec{b}_{i_k}) / |\vec{a}_{i_k} \cdot \vec{b}_{i_k}|$$

where

$$\vec{a}_{i_k} = (x_i - x_{i+k}, y_i - y_{i+k})$$

$$\vec{b}_{i_k} = (x_i - x_{i-k}, y_i - y_{i-k})$$

A point is marked as an angle if both:

1) $C_k(i) > t$, where t is some predefined threshold, and

2) $C_k(i)$ is a <u>local maxima</u>, i.e., $C_k(i) \geq C_k(i+\Delta)$, $|\Delta| \leq d$, where d is a predefined radius for the computation of local maxima. Once the angles are detected, consecutive angles can be connected by straight lines to form the polygonal approximation.

The first functional approximation procedure is due to Ramer (1972) and is based on successively splitting any segment which is not well approximated by a straight line joining its end points. The measure of goodness of fit used is the greatest distance of any point in the segment from the approximating line. If this distance is too great, then that segment is split into two segments at the point of maximal departure.

Ramer's algorithm can only split segments into smaller segments; it has no facility for merging two segments into one longer segment. A procedure which admits both splitting and merging is described by Pavlidis (1977).

An unfortunate aspect of all of these polygonal approximation procedures is that even though the input curve may be non-self-intersecting, there is no guarantee that the resulting approximation will not self-intersect. The post-editing that might be required to check and correct for such intersections can add significant overhead to computing the approximation.

3.2 Matching Based on Polygonal Approximations

In this section we present a shape matching procedure based on a polygonal approximation representation of a shape. This procedure is invariant to rotation and a wide range of scale changes; it will construct and manipulate a structure called an association graph. An important property of this approach is its ability to deal with incomplete information.

Let $O = O_1, O_2, \ldots, O_n$ be the polygonal path representation of some object curve in a database of curves. Each O_i is an angle represented by an ordered triple, $\langle X_i, Y_i, MAG_i \rangle$, where X and Y are the coordinates of O_i, and MAG_i is the magnitude of the angle. Similarly, let $T = T_1, T_2, \ldots T_m$, $m < n$, be the polygonal representation of an unknown curve segment that we wish to match against O. Informally, a match of T to O has two components:

1) a function providing an association of elements in T with object elements, called an association function, and

2) a global coordinate transformation from the coordinates of T to the coordinates of O that provides a spatial registration of T with O.

The first step in matching T against O is to compare the angles in T against the angles in O using a <u>local evaluation function</u>. The local evaluation function will determine, for each angle in T, a subset of object angles that the angle in T may be associated with. Each such association of an angle in T with a single matching angle in O will correspond to a node in a graph, called the association graph. Each node will be assigned a <u>weight</u> reflecting how well the two angles match according to the local evaluation function.

Connections between nodes are determined as follows: Denote a node in the graph by $<0_i,T_j>$. Suppose $<0_i,T_j>$, $<0_{i'},T_{j'}>$ are two such nodes, and suppose that prior knowledge is available concerning the scale of 0 relative to T. Then the two nodes can be connected if $d(0_i,0_{i'})/d(t_j,T_{j'})$ is consistent with our prior knowledge of scale (here $d(0_i,0_{i'})$ is the distance between 0_i and $0_{i'}$). If no such information is available, then all pairs of nodes are connected.

The graph is next operated on by two processes -- one which examines (and possibly deletes) nodes, and the other which examines edges. The node process will prune a node from the graph if some evaluation function applied to the neighborhood of the node is below threshold. For example, consider the following evaluation function. Let $G = <N,V>$ be the association graph with node set N and arc set V. Suppose $M(n)$, for $n \varepsilon N$, is the weight, or merit, of node n. Then the function

$$E(n) = \sum_{\substack{n' \text{ s.t.} \\ (n,n')\varepsilon V}} M(n')$$

simply sums the weights of all the neighbors of node n. The node deletion process using a particular threshold t on $E(n)$ will define a sequence of association graphs $G = G_0, G_1, \ldots, G_s$, where G_i is obtained by G_{i-1} as follows:

1) $N_i = N_{i-1} - \{n: n\varepsilon N_i \text{ and } E(n) < t\}$

2) $V_i = \{(n,n') \mid n,n' \varepsilon N_i\} \cap V_{i-1}$

When $G_i = G_{i-1}$, the process terminates. We will let G^∞ denote the fixed point of this process.

Proposition: For some finite s, $G_s = G^\infty$.

Proof: Immediate, since G_0 is finite, and each iteration which does not reach the fixed point reduces the size of the network (and the empty network is a fixed point).

Note that if no prior scale information is available, then, since all pairs of nodes are connected (i.e., the graph is complete), the node deletion process will not help at all in establishing a best match. However, by attaching information to the edges of the graph, and introducing an edge pruning process we can significantly enhance the power of the node pruning process since pruning an edge will change the neighborhood of nodes at the end of that edge, which may, in turn, cause them (and subsequently other nodes and edges) to be pruned.

An edge in the association graph connects two associations $<0_i,T_j>$ and $<0_{i'},T_{j'}>$. The similarity transformation that maps the line segment joining T_j to $T_{j'}$ onto the line segment joining 0_i to $0_{i'}$ is unique, and corresponds to a local hypothesis of the best coordinate transformation that minimizes the mean squared distance between T and 0 given that $<0_i,T_j>$ and $<0_{i'}, T_{j'}>$ are part of the best overall association. The parameters of this transformation will be associated with the edge joining $<0_i,T_j>$ and $<0_{i'},T_{j'}>$. If these associations are in fact part of the best association function from T to 0, then many other edges in the network should be labeled with similar parameters. In particular, these transformations should be found attached to the other edges that are incident on $<0_i,T_j>$ and $<0_{i'},T_{j'}>$.

This suggests that the following new graph (the line graph), G', be built from the given association graph, G.

Definition: Let $G = \langle N,V\rangle$ be an undirected graph with node set N and edge set $V \subseteq N \times N$. Then the line graph, G', of G is the graph (N',V') where

1) for each $v \in V$ we have a node $n' \in N'$;

2) $(n_1',n_2') \in V'$ iff n_1',n_2' correspond to edges in V that have one common endpoint.

Then, a node can be pruned from this graph, if, say, the value of some evaluation function applied to that node is less than some threshold. This can be done in the following way. Let $n_1',n_2' \in N'$. Suppose n_1 corresponds to the edge joining $\langle 0_{i1},T_{j1}\rangle$ to $\langle 0_{i2},T_{j2}\rangle$ in the original association graph, G. Then the similarity transformation labeling n_1' is the 4-tuple (a,b,c,d) where

$$\sum_{r=1}^{2} |X_{ir} - (aX_{jr} + bY_{jr} + c)|^2 + |Y_{ir} - (aY_{jr} + bX_{jr} + d)|^2 = 0$$

Then node n_2' must correspond to an edge in G connecting some $\langle 0_{i3},T_{j3}\rangle$ to either $\langle 0_{i1},T_{j1}\rangle$ or $\langle 0_{i2},T_{j2}\rangle$. Suppose it is $\langle 0_{i1},T_{j1}\rangle$. If both n_1' and n_2' were part of the same perfect match of T to 0, then it would be the case that

$$s(n_1'|n_2') = \sum_{r=1,3} |X_{ir} - (aX_{jr} + bY_{jr} + c)|^2 + |Y_{ir} - (aY_{jr} + bX_{jr} + d)|^2 = 0$$

If the match were not perfect, or if n_1' and n_2' were not part of a single match, then, in general, $s(n_2'|n_1') > 0$.

We can use the evaluation function

$$E'(n') = -\sum_{n''} s(n''|n') \quad \text{with } (n',n'') \in V'$$

to create a sequence of graphs $G' = G_0', G_1', \ldots, G_s'$, where G_1' is related to G_{i-1}' by

1) $N_i' = N_{i-1}' - \{n' \varepsilon N_{i-1}' | E'(n') < t\}$;
2) $V_i' = \{(n', n'') | n', n'' \varepsilon N_i'\} \cap V_{i-1}'$.

By introducing the line graph, the same system of programs that performed node deletion can now be applied to the edge deletion process. Also note that if G' is the fixed point of this process, then it is reached after a finite number of steps.

It is important to note that a more efficient implementation of the pruning algorithms is possible than the one implied by the definitions of the processes, which involves examining each node of G_i to produce G_{i+1}. The more efficient algorithm is motivated by the observation that deleting a node from G_i only changes the neighborhoods of the neighbors of those nodes. Therefore, only the neighbors of those nodes deleted from G_i (and which are also included in G_{i+1}) need to be examined to compute G_{i+2}.

If the association graph is examined after both pruning operations have been applied to it, it will usually be the case that it contains at most a few connected components of nodes with highly consistent similarity transformations attached to the edges of these components. Searching for a best association function should almost always be trivial. We have also constructed powerful clues as to the best global similarity transformation between T and O.

3.3 Searching the Pruned Association Graph

This section discusses methods for extracting a "best" association function from the filtered graph. These techniques would be applicable to the original graph. However, since they have a sequential component, it would be computationally more costly to directly process the unfiltered graph.

Let G_f represent the final association graph produced by the node and edge deletion processes; G_f represents a reduced and more informed set of guesses as to which 0_i should be associated with a particular T_j. We will discuss the use of ordered search techniques to find the best association function contained in G_f. The combination of state-space search and "discrete" relaxation processes is very similar to the incorporation of probabilistic relaxation with heuristic search in MSYS (Barrow and Tenenbaum 1976).

The states in the search space will be described by association functions and contexts (see below) and the search process will consist of choosing the state S with minimal cost, and creating two new states.

Given a state, S, let F_s denote the state association function for S and let C_s denote the context for S.

F_s is a set of ordered pairs

$$\{\langle 0_{i1},T_{j1}\rangle, \langle 0_{i2},T_{j2}\rangle, \ldots, \langle 0_{in},T_{jn}\rangle\}$$

such that:

1) $0_{ip} = 0_{iq} \Rightarrow p = q$
2) $T_{jp} = T_{jq} \Rightarrow p = q$

It represents the partial development of some association function from T into 0.

Each C_s is a connected subgraph of G_f. The nodes of C_s represent the possible associations which may be used to construct extensions to F_s.

We will now describe how a state S is expanded to produce two new states, S' and S''. Let $<0_i,T_j>$ be in C_s but not in F_s. Then, S' is obtained by adding $<0_i,T_j>$ to F_s to obtain $F_{s'}$. S'' is obtained by deleting $<0_i,T_j>$ from C_s to obtain $C_{s''}$. In other words, S' corresponds to the assertion that $<0_i,T_j>$ *is* part of the best match of T against 0 constrained to include F_s, while S'' corresponds to the assertion that it is *not*.

Considering S', since $<0_i,T_j>$ is in F_s, it is clearly unnecessary to retain any node which associates 0_i with any $T_{j'}, j' \neq j$, or T_j with any $0_{i'}, i' \neq i$. Such nodes can therefore be deleted from C_s in forming $C_{s'}$. Note, however, that pruning these nodes changes the neighborhoods of other nodes, so that the node and edge pruning processes can be reapplied. This is accomplished by the following algorithm (Algorithm H-Hypothesize):

1) Set $C^0_{s'} = C_s - \{<0_{i'},T_j> | i' \neq i\} \cup \{<0_i,T_{j'}> | j' \neq j\}$, i.e., delete from C_s any associations involving either 0_i or T_j except $<0_i,T_j>$.

2) Apply the node and edge deletion procedures iteratively to $C^0_{s'}$ to obtain $C^1_{s'}, C^2_{s'}, \ldots, C^k_{s'}$, where $C^k_{s'}$ is the fixed point. Set $C_{s'} = C^k_{s'}$, provisionally.

There are logically three relations that $F_{s'}$ might have to $C_{s'}$:

1) Some of the associations contained in $F_{s'}$ are deleted

by the pruning processes. This indicates that $F_{S'}$ was constructed from incompatible associations, and does not describe the best association function.

2) $C_{S'}$ is disconnected by the pruning processes, and the associations of $F_{S'}$ are split among the various connected components of the $C_{S'}$. Once again, $F_{S'}$ was constructed from incompatible associations.

3) The associations of $F_{S'}$ are all contained in a single connected component of the $C_{S'}$. In this case, that connected component becomes $C_{S'}$.

The only allowable extensions will be those that fall in the third category. The association functions for such states will be called constraint compatible. Any state S that cannot be extended to an S' because all attempts at extension result in one of the first two conditions will be marked with very high cost. The association functions for such states will be called constraint incompatible. In this way, these states need not be considered during the remainder of the search process.

Next consider the generation of S''. $F_{S''} = F_S$, but we are deleting $<0_i,T_j>$ from C_S. Again, this can change the neighborhoods of nodes in C_S, so that the final value of $C_{S''}$ is obtained by the following process (Algorithm S-Suppress):

1) Set $C^0_{S''} = C_S - \{<0_i,T_j>\}$

2) Iteratively apply the node and edge deletion processes to obtain the sequence of graphs $C^1_{S''}, C^2_{S''}, \ldots, C^k_{S''}$, with $C^k_{S''}$ the fixed point. Set $C_{S''} = C^k_{S''}$.

In summary, the state-space contains states S that are ordered pairs (F_S,C_S). The start space is $(\emptyset,G)$ and a final state is a pair (F_S,C_S) where the nodes in C_S are identical to the associations in F_s (i.e., there is no way to expand F_s). Given some cost function and a method for computing a lower bound estimate for each state of the cost of a minimal cost association function constrained to include that state, we can describe the search algorithm as follows (Algorithm F):

(0) Put (Ø,G) on the list of OPEN states.

(1) Choose that state S from OPEN that has minimal cost estimate. If C_s contains only the associations in F_s, stop with F_s being the optimal association function.

(2) Pick some $<0_i,T_j>$ from C_s to build S' and S''

(a) $F_{s'} = F_s \cup <0_i,T_j>$, $F_{s''} = F_s$

(b) Obtain $C_{s'}$ by Algorithm H

(c) Obtain $C_{s''}$ by Algorithm S

(d) Put S' and S'' on OPEN.

(3) Start again at step (1).

Ordered search techniques can be used both to direct search towards the best terminal state and also to terminate search along a path that could not conceivably be optimal. See Davis (1979) for a discussion of a general cost function and cost estimating procedure which can be used to guide the search.

3.4 Experimental Study

A matching experiment was performed using the coastlines of several islands as the input shapes. The source of the data was World Data Bank I, a digital representation of the coastlines and boundaries of the world specifically designed for automated mapping systems (WDB). World Data Bank I is in the form of (longitude, latitude) coordinates. The coastlines were transformed to Cartesian coordinates using the Lambert normal conical projection based on a standard parallel. The data base of objects was created using one set of standard parallels, while the unknowns were produced using a different set of standard parallels. This introduced systematic distortions between the objects and unknowns.

After the projection, each individual coastline was scaled to fit into a 127x127 array so that gross differences in the physical sizes of the coastlines would not serve as a cue for the matching process.

Five coastlines were chosen -- Baffin Island, Cape Breton Island, Cuba, the Dominican Republic, and Guadeloupe. The unknowns were segments chosen from the same coastlines, but projected at different latitudes. Polygonal path representations for the objects and the unknowns were constructed using the angle detection procedure described previously.

In an attempt to provide a difficult matching environment for the matching processes, the unknown shapes were degraded according to the following procedure: Given shape T = $\{(X_i,Y_i)\}_{i=1}^{m}$, choose an arbitrary scale factor, a, and a noise factor, q. Each element of T is multiplied by a + s, where s is chosen from the normal distribution with mean 0 and variance q. That is, we construct a $T' = \{(X_i',Y_i')\}_{i=1}^{m}$ where

$$X_i' = X_i(a + s), \text{ and}$$
$$Y_i' = Y_i(a + s)$$

Nodes in the association graph were connected if the associated distances, d_1 (for the template elements) and d_2 (for the object elements), satisfied $.4 < ABS(d_1 - d_2)/d_1 < 2.5$, allowing for a wide range of scale changes. The values of these thresholds determine the degree of scale invariance of the procedures. It should be pointed out that allowing such a wide range of scale changes will clearly place most of the burden on the edge deletion process, and that the node deletion process will only be able to delete the grossest mismatches.

The node deletion criterion was to delete a node when the sum of the weights on its neighbors was less than a threshold, t_d. If the unknown were an exact match to a segment of one of the objects, then we would use a threshold of $t_d = (m - 1)c$, (c = minimum value of local angle matches allowed for introducing a node into the network), since every correct association O_i, T_j would be connected to m - 1 other correct associations. However, since we do not have exact matches, but only approximate matches, we have adopted the following rule which was used consistently: set t_d equal to 2/5 of the number of prominent angles in the unknown shape, where a prominent angle was taken to be one with a magnitude less than 60 degrees. In this way, we allow for matches that only approximately account for the most salient features.

The edge deletion process is defined as follows: Consider the triple of nodes shown in Figure 3, and suppose we are

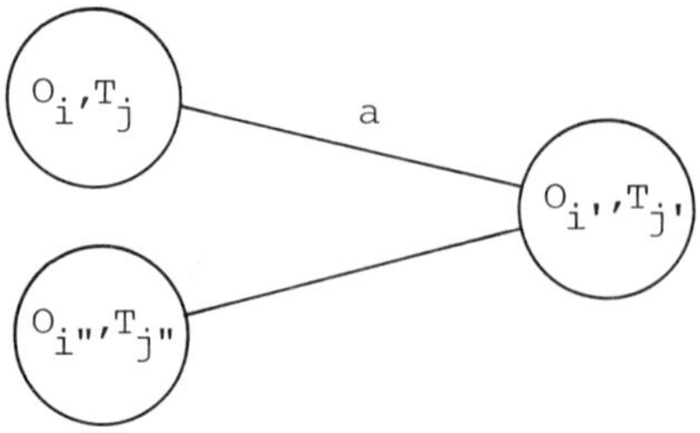

FIGURE 3. Edge Deletion

evaluating edge a. The similarity transformation associated with edge a is applied to $0_{i''}$ and the distance between the transformed $0_{i''}$ and $T_{j''}$ is computed. If this distance is less than some threshold, one piece of evidence is counted for edge a.

In 47 out of the 50 match trials, the edge deletion process deleted all mismatch associations, while preserving all of the match associations, thus eliminating the need to search the association graph. In the remaining three cases where this situation did not occur, only nominal searching was required (see Davis 1979 for more details).

4. CONCLUSIONS

This paper has discussed shape matching algorithms, initially developed for computer pattern analysis, which are applicable to matching cartographic data. There are, however, several outstanding problems which must be addressed before truly useful shape matching facilities can be realized:

1) The database problem -- as discussed in the introduction, not much effort has been devoted to the problem of organizing large numbers of structurally modeled shapes into a database. Some work has been done on databases for quadtree representations (Davis and Roussopoulos 1979), but these suffer from certain deficiencies as far as shape matching is concerned.

2) The mismatch description problem -- it is unlikely that the goal of any cartographic shape analysis system will be to simply find the best match to an unknown shape. Rather, it will ordinarily be required to describe how the shape from the database differs from the unknown shape. Although the

exact form of those descriptions will be problem dependent, it is still true that they must be inferred from some general purpose vocabulary for describing shape differences and similarity. It is unlikely that either chain codes or polygonal approximations can be used for these purposes. Such a vocabulary should _minimally_ be able to describe shapes in terms of protrusions, intrusions, symmetries and prominent axes. Much more work needs to be done in this area.

3) We have been exclusively concerned with two-dimensional shapes in this paper. However, cartographic data often has higher dimensionality (e.g., digital terrain models). The problems involved in matching higher dimensional structures, or projections of such structures, against one another, is a problem currently being faced by computer scientists, and one which may be faced in cartographic applications.

REFERENCES

Barrow, H., and Tenenbaum, J. M. (1976). "MSYS: A System for Reasoning About Scenes," SRI AI Center TN-121.

Davis, L. (1976). A Survey of Edge Detection Techniques, _Computer Graphics and Image Processing_ _4_, 248-270.

Davis, L. (1979). Shape Matching Using Relaxation Techniques, _IEEE Trans. Pattern Analysis and Machine Intelligence_ _1_, 60-72.

Davis, L., and Roussopoulos, N. (1979). "Approximate Pattern Matching in a Pattern Database System," University of Texas, Computer Sciences Department TR-99.

Dudani, S., Breeding, K., and McGhee, R. (1977). Aircraft Identification by Moment Invariants, _IEEE Trnas. Computers_ _26_, 39-45.

Feder, J., and Freeman, H. (1965). "Segment Fitting of Curves in Pattern Analysis Using Chain Data," AD 619 525.

Freeman, H. (1961). On the Encoding of Arbitrary Geometric Figures, IEEE Trans. Elec. Comput. 10, 260-268.

Pavlidis, T. (1977). "Structural Pattern Recognition," Springer-Verlag, Berlin.

Persoon, E., and Fu, K. S. (1977). Shape Discrimination Using Fourier Descriptors, IEEE Trans. Systems, Man and Cybernetics 7, 170-179.

Ramer, U. (1972). An Interactive Procedure for the Polygonal Approximation of Plane Curves, Computer Graphics and Image Processing 1, 244-256.

Rosenfeld, A., and Kak, A. (1976). "Digital Picture Processing," Academic Press, New York.

World Data Bank I, NTIS (Magnetic Tape) PB-233 178/PDK.

Zucker, S. (1976). Region Growing: Childhood and Adolescence, Computer Graphics and Image Processing 5, 382-399.

THE EFFECTS OF GENERALIZATION IN GEOGRAPHICAL DATA ENCODING

Michael F. Goodchild

Department of Geography
University of Western Ontario
London, Canada

Any digital map data will differ from the real world it represents, usually as the result of a number of forms of generalization. An understanding of the effects of generalization on the accuracy of digital data is essential to effective system design. This paper reviews approaches to the estimation of data errors for both raster and vector systems. In the vector case the discussion centres on the spurious polygon problem.

I. INTRODUCTION

Generalization can occur at a number of stages in the transfer of data from the real world to a digital image. In some cases it is subject to precise rules, as in the encoding of a continuous spatial variation of electromagnetic radiation by a finite number of pixels. In others the process is subjective, as for example in the smoothing of contour and boundary lines by a cartographer working from aerial photography, or by a digitizer operator following a line. It is inevitable that information will be lost in one or more of the steps between reality and encoded data.

This paper is concerned with the effects of generalization error on the subsequent use of digital map data. The most useful products of a geographical information system, or more generally any form of map data processing, are measures of

ISBN 0-12-267180-5

some kind, such as the area of a homogeneous patch of land of certain characteristics, the length of a line or the distance between specified points. In both raster and vector encoding the accuracy of these products, with respect to the true, real world values, is a function of the generalization inherent in the capture process. The greater the generalization, the lower the accuracy. But at the same time, greater generalization usually implies a smaller volume of data, and faster processing. So an understanding of the relationship between accuracy and generalization is essential to effective system design.

The paper is divided into two sections. The first deals with raster encoding, and the effects of the generalization implicit in the choice of raster or pixel size on the products of a raster-based system. The second deals with a generalization problem normally associated with vector data. In the latter case generalization is usually considered to be a function of the density of sample points along each line.

The accuracy problem would be simple if measurements from digital data could be checked directly against their true, real world values. But this is not normally possible. In general accuracy must be predicted from the digital data alone, by making assumptions about the true data. Clearly accuracy estimates will be most useful when based on the least restrictive and most realistic assumptions.

II. RASTER ACCURACY

There are two significant estimation problems in the raster context:

i) Area: what are the confidence limits on an estimate of the area of a patch of land with a particular set of characteristics, when the patch is represented in raster form?

ii) Point: what is the probability of error in estimating the characteristics of a point from a raster?

A. Area Estimation

The first problem has received the widest recognition because of its relationship to a standard method of manual measurement of area used in forestry and photogrammetry and known as dot planimetry. A transparent sheet covered with a square array of dots is laid over the patch to be measured, and the number of dots lying within the patch is used as the basis of an area estimate.

A simple error model can be constructed by assuming that each dot has an equal and independent probability of lying within the patch. The dot count would then have a binomial distribution. For a constant area and variable dot spacing, the standard error of the area estimate would depend on the dot count N to the negative one half power (1).

However the binomial model is clearly unrealistic for most geographical data, except when patches are very small and fragmented with respect to the raster cell size. The area to be measured usually consists of one or more patches each extending over a number of contiguous raster cells. All of the measurement error is then attributable to those cells on the edges of each patch, invalidating the binomial model. Some of the boundary cells will be counted as part of a patch, but contain land which is in reality not part of the patch; these contribute to an overestimation of area. Conversely there are cells which are not counted, and yet overlap the patch; these contribute to underestimation.

Frolov and Maling (2) estimated the contribution of each boundary cell by assuming that the boundary could be regarded as a straight line drawn randomly across the cell. They found the mean square area of the cutoff portion, that is, the error variance in the area estimate for each boundary cell, to be αb^4, where b is the linear dimension of the cell, assumed to be square, and $\alpha = 0.0452$. Lloyd (3) pointed to a problem with the process used by Frolov and Maling to define a random straight line, and recomputed α as 0.0609. In an independent paper Goodchild and Moy (4) obtained 0.0619 by integrating the same expression (compare Appendix 1 of (4), p. 79, with (3), p. 25).

The error variance in the estimate of area depends on the summation of the individual errors from each of the boundary cells. Let n represent the number of cells intersected by the boundary. Then assuming the contributions of each cell to be independent, the error variance will be $n\alpha b^4$, giving a standard error of $(n\alpha)^{1/2}b^2$. Bellhouse (5) has analyzed the case where the independence assumption is invalid, and see also (6).

Frolov and Maling estimated the value of n by relating it to the perimeter length of the patch. For a constant patch, we can expect that a halving of the cell dimension b will lead to a quadrupling of N, the number of cells counted, and a doubling of the length of the perimeter and n. Thus n is proportional to $N^{1/2}$, and we can write the standard error of estimate as $kN^{1/4}\alpha^{1/2}b^2$. Since the estimate of area S is Nb^2, the standard error as a percentage of the estimate will be proportional to $N^{-3/4}$

$$\varepsilon = k\alpha^{1/2}N^{-3/4} \qquad (1)$$

Alternatively, if we regard the variable as the cell size, b, the dependence of percentage error on b is to the power 1.5. A number of studies have verified these relationships empirically (see (4), (7) and a reanalysis of Tobler's data by Goodchild (6)).

The constant of proportionality k depends on the shape of the patch. A long, thin patch has more boundary cells than a circular patch of the same area, and hence a greater standard error. Frolov and Maling (2) computed values of k for various standard shapes, again under the independent straight line assumption. Moy (7) examined the hypothesis that the population of patches present on a map sheet of a certain type could be regarded as characterized by a single value of k.

B. Point Estimation

The point estimation problem is to determine the probability that the characteristics of some chosen point, as determined from a raster representation, are correct. For a single patch the probability π_1 that a point in the patch will be misclassified as belonging to some other, neighbouring patch is given by the total of the overestimating portions of boundary cells as a percentage of patch area. Similarly the probability π_2 that a point not in the patch will be misclassified as belonging to it is derived from the total of the underestimating portions.

The relationship between these probabilities and the previous analysis is straightforward. For those boundary cells which contribute to an overestimation of area, let δ

denote the mean fraction of cell area contributed. One half of the boundary cells are expected to contribute to overestimation, giving a total area of $n\delta b^2/2$. Thus $\pi_1 = n\delta b^2/2S$. Using the same argument for those cells contributing to underestimation and substituting for n gives

$$\pi_1 = \pi_2 = k^2 \delta b S^{-1/2}/2 = k^2 \delta N^{-1/2}/2 \qquad (2)$$

Misclassification probabilities are thus inversely proportional to the square root of the number of cells counted. Moy (7) found that empirical evidence supported this analysis.

III. FRACTIONAL DIMENSIONALITY

In assuming n to be proportional to $N^{1/2}$ we assume in effect that boundaries are smooth. For contorted lines, or for cells which are large compared to the patches they represent, we can expect n to increase more rapidly. There are strong parallels here with work on the relationship between the length of geographic lines and the scale at which length is measured. Richardson (8) showed that for many real lines, of which the west coast of Britain has become the classic example, the relationship between length and scale is remarkably regular. Thus if length is measured by stepping a pair of dividers along such a line, the length estimate varies with the sampling interval, in other words, the dividers setting, raised to a negative power. The power will be zero and the length estimate constant only in the case of a smooth curve.

Mandelbrot (9) has placed Richardson's empirical results within the concept of fractional dimensionality. A curve is said to be a fractal if its length varies with the sampling interval in the manner observed by Richardson, and is said to have dimension D if the relationship is of power 1-D. D is therefore 1 for a smooth curve, and has a theoretical limit of 2 in the case of a line so contorted that it fills the area.

In the case of a raster representation of a line, the size of the raster b corresponds to a sampling interval. Thus we expect n, the number of cells intersected by the line, to depend on b^{-D}. For a constant area, n depends on the number of cells counted to the power D/2, which gives the result used previously, $n = k^2 N^{1/2}$, only for a smooth curve.

Fractional dimensionality thus provides a way of characterizing the dependence ε, π_1, and π_2 on raster size when patch boundaries are appreciably wiggly. The revised estimates in terms of cell size are as follows (6)

$$\varepsilon = k\alpha^{1/2}\, b^{2-D/2}/S \tag{3}$$

$$\pi_1 = \pi_2 = k^2 \delta\, b^{2-D}/S \tag{4}$$

IV. SWITZER'S ANALYSIS

Switzer (10) presented an entirely different approach to the problem of point estimation, in terms of 'mismatch area'. The area assigned to a patch by a representative raster but in reality lying outside the patch is equal to $\pi_2 S$; similarly $\pi_1 S$ is the area misrepresented as lying in the patch. In the following the patch area is referred to as black and the surrounding area as white; counted raster cells are similarly denoted as black. Write $P_2(d)$ as the probability that a randomly chosen point in a black cell, distance d from the centre of the cell, is in fact white. Then we have:

$$\pi_2 = S^{-1} \sum_{h=1}^{N} \int_{A_h} P_2(|\underset{\sim}{s} - \underset{\sim}{s}_h|)\, d\mu(\underset{\sim}{s}) \tag{5}$$

where the summation is over each of the N black cells. The integral is over the area of cell h, and $\underset{\sim}{s}_h$ denotes the location of the central point in the cell. $|\underset{\sim}{s} - \underset{\sim}{s}_h|$ is therefore the distance between the centre and every other location.

Switzer proposed that $P_2(d)$ be approximated by a Taylor expansion in its derivatives with respect to d. Taking the first two terms, and recognizing that $P_2(0) = 0$ we have

$$P_2(d) = P_2' \, d + \frac{1}{2} P_2'' \, d^2 \qquad (6)$$

where P_2' and P_2'' are the first two derivatives evaluated at $d = 0$. Thus

$$\pi_2 = [P_2' \, \rho_1 + \frac{1}{2} P_2'' \, \rho_2] \qquad (7)$$

where ρ_1 and ρ_2 are the mean and mean square distances of a random point from the cell centre, respectively.

P_2' and P_2'' can be estimated from the pattern of white and black cells in the following way. Let m(1) denote the number of times a black cell can be found with a white 4-neighbour. Similarly let m(2) denote the number of times a black cell has a white 4-neighbour two cells away. A good estimate of π_2 is then given by

$$\pi_2 = [0.60 \, m(1) - 0.11 \, m(2)]/4N \qquad (8)$$

Empirical work by both Switzer and Muller (11) has supported this analysis.

The relationship between m(1), m(2) and scale can be predicted from the fractional dimension D of the boundary of the patch as follows (6). Consider one row of cells. For a given area, the number of black cells in the row is proportional to b^{-1}, and the number with white neighbours to b^{1-D} (12). Thus if the contour has dimension 1, the number of cells with white neighbours is a constant independent of the number of black cells, but for $D > 1$ it rises with increasing resolution. Since the number of rows containing black cells is proportional to b^{-1}, the total count of black cells with white neighbours, m(1) is proportional to b^{-D}. Similarly m(2) depends on $(2b)^{-D}$. In general Switzer's estimate can be rewritten as

$$\pi_2 = K(D)\ S^{-1}\ b^{2-D} \tag{9}$$

K(D) is given by $K'[0.60 - 0.11\ (0.25)^{D/2}]/4$ where K' is the constant of proportionality in the equation $m(1) = K'\ b^{2-D}$. Switzer's result is thus compatible with the previous analysis for π_2, in the dependence of π_2 both on cell size b, and on the dimensionality of the patch boundary.

The Switzer and Frolov and Maling approaches are very different in application. Switzer's estimate of mismatch area is obtained from the grid representation itself. Frolov and Maling's estimate of standard error is based on assumed characteristics of the unobservable 'true' map, since k must be obtained from the patch shape.

V. VECTOR ACCURACY AND SPURIOUS POLYGONS

The determinant of generalization in a vector representation is the density of sample points along a line. If the data has been encoded by vectorization of a raster scan, the sampling density will depend on the raster size. If the data has been digitized by line following, the sampling density may have been determined implicitly by the operator, or may have been established as a system parameter.

The effect of sampling density on the accuracy of area and point estimates is obtained by a similar argument to that in the case of rasters above, and can also be related to fractional dimensionality. In particular, the expressions for the standard error of area estimate and misclassification probability have the same dependence on the mean interval between sample points as they do on cell size in the raster case.

Particular problems arise in the vector case as a result of generalization when several coverages are overlaid. Figure 1 shows an example in which census divisions for two different time periods have been overlaid. In some cases boundaries on the two maps are roughly coincident over long stretches because both represent independent samplings of the same real geographical line.

FIGURE 1. Overlay of Census Divisions from two time periods for British Columbia.

TABLE I: Polygons by Area for Five CGIS Coverages and Overlays

Acreage	924C	925C	926C	927C	928C	UTES	UTE4	URIS
0-1	0	0	0	1	2	2640	27566	77346
1-5	0	165	182	131	31	2195	7521	7330
5-10	5	498	515	408	10	1421	2108	2201
10-25	1	784	775	688	38	1590	2106	2129
25-50	4	353	373	382	61	801	853	827
50-100	9	238	249	232	64	462	462	413
100-200	12	155	152	158	72	248	208	197
200-500	21	71	83	89	92	133	105	99
500-1000	9	32	31	33	56	39	34	34
1000-5000	19	25	27	21	50	27	24	22
>5000	8	6	7	6	11	2	1	1
Total	88	2327	2394	2149	487	9558	39188	90599

924C Recreation Capability

925C Land Use, 1964

926C Land Use, 1968

927C Land Use, 1973

928C Soil Capability for Agriculture

UTES 924C + 926C + 928C

UTE4 UTES + 927C

URIS UTE4 + 925C

Note: 1 acre = 0.405 hectare

The number of polygons produced in an overlay depends not so much on the number of polygons overlaid, but on the complexity of each boundary. Two polygons of n_1 and n_2 vertices can produce from 3 to $n_1 n_2 + 2$ overlay polygons, classified into 16 logical combinations (13). Frequently both maps in an overlay will contain digital representations of the same real line. Since the maps have usually been subjected to independent generalization, the overlay will contain large numbers of 'spurious' polygons. Table I shows the result of overlaying five coverages in the Canada Geographic Information System; the area covered is the Ottawa-Hull National Capital Region. An enormous number of spurious polygons has been produced principally because of the persistence of boundary lines in the three land use coverages. Even though they occupy a small proportion of total map area, large numbers of spurious polygons and associated arcs create serious problems for topological data structures and should be suppressed, provided that an effective means can be found for discriminating between spurious polygons and real ones. Paradoxically, the more accurately the digitization and the greater the density of sample points, the greater the number of spurious polygons produced. Generalization of boundary lines tends to reduce the spurious polygon problem rather than increase it.

Goodchild (13) has argued that all forms of vector data capture tend to produce close to the maximum possible number of spurious polygons. For two polygons of n_1 and n_2 vertices the maximum is $2 \min(n_1, n_2) - 4$, which is substantially greater than the $2n_1 n_2/(n_1 + n_2) - 3$ expected when sample points are randomly located along each digitized line. Empirical evidence tends to support this.

Size is the simplest readily available criterion for suppressing spurious polygons, but other criteria are available which might be used to improve the accuracy of deletion. All spurious polygons have only two arcs. In addition the presence of one spurious polygon makes neighbouring polygons created by the same pair of original arcs more likely to be spurious. Finally, spurious polygons tend to have a distinctive shape. These possibilities are being explored in current research.

VI. CONCLUSIONS

Accuracy has received almost no attention as a design criterion in geographical data processing, although accuracy is always lost at the capture stage, and may also be affected later during processing. There are two reasons for this. First, design has been dominated until recently by technological issues, and accuracy assumed to be outside the designer's control. Recently technical issues have tended to become less important, and at the same time applications have begun to reach the level where the demonstration of capability is no longer sufficient: users demand accuracy and performance as well. Second, there is little literature on map generalization, which has always been treated implicitly, whereas the use of computer processing requires that such issues be addressed as explicitly as possible.

The accuracy of many systems is poor, particularly in the natural resource field, to the extent that system products often have little utility. Accuracy also tends to deteriorate through time, and few systems have been designed with effective methods for updating, since use levels can rarely justify this. If geographical data processing is to advance from a research and demonstration mode to real utility, it is essential that far greater emphasis be placed on evaluating accuracy and error in both design and operation.

REFERENCES

1. Tobler, W. R., "The Accuracy of Categorical Maps." Department of Geography, University of Michigan. Cartographic Laboratory Report Number 4 (1974).
2. Frolov, Y. S. and D. H. Maling, *Cart. J. 6*, 21 (1969).
3. Lloyd, P. R., *Cart. J. 13*, 22 (1976).
4. Goodchild, M. F. and W. S. Moy, *in* "Proceedings of the Commission on Geographical Data Sensing and Processing, Moscow, 1976" (R. F. Tomlinson, ed.), International Geographical Union, Ottawa (1977).
5. Bellhouse, D. R., "Area Estimation by Point Counting Techniques." Statistics and Actuarial Science Group, University of Western Ontario (1979).
6. Goodchild, M. F., "Fractals and the Accuracy of Geographical Measures." Department of Geography, University of Western Ontario (1979).
7. Moy, W. S., "Estimation from Grid Data: the Map as a Stochastic Process." Unpublished M.A. Dissertation, Department of Geography, University of Western Ontario (1977).
8. Richardson, L. F., *Gen. Syst. Yearbook 6*, 139 (1961).
9. Mandelbrot, B. B., "Fractals: Form, Chance and Dimension." Freeman, San Francisco (1977).

10. Switzer, P., *in* "Display and Analysis of Spatial Data" (J. C. Davis and M. J. McCullagh, eds.), Wiley, London (1975).

11. Muller, J.-C., *Can. Cart. 14*, 152 (1978).

12. Mandelbrot, B. B., *J. Fluid Mech. 72*, 401 (1975).

13. Goodchild, M. F., *Harvard Papers in Geographic Information Systems 6* (1978).

METHODOLOGICAL OBSERVATIONS ON THE STATE OF GEOCARTOGRAPHIC ANALYSIS IN THE CONTEXT OF AUTOMATED SPATIAL INFORMATION SYSTEMS

Avi Degani

Department of Geography
Tel Aviv University
Tel Aviv, Israel

A Geographer is a person who does not know where Ecaf is (or Chicago), but can tell you everything about why it is there....

A Computer Specialist is a person who does not know what spatial analysis is all about, but can offer you a faster algorithm to execute a mapping solution for it
or/and compile for it the necessary data...
and/or classify these data methodologically...
and/or store these data on lesser magnetic tape space than you ever expected...

THE TOWER OF BABYLON

To counter geographers' old joke about their own doing, it seems in order to raise a brow concerning some recently prevailing trends among computer scientists whereby sophisticated solutions are suggested to non existing or least relevant, map data processing problems. By definition data processing, is the <u>spatial</u> processing of data, or the processing of <u>spatial</u> data, hence a geographic analysis. Naturally,

ISBN 0-12-267180-5

he logical focus of this activity must be the spatial problem, not the tools nor the means employed in the analysis - the computer environment in our case. Embarrassingly, there seems to be an increasing evidence of a deminishing meaningful communication between those scholars who deal with analytical cartography aimed at solving spatical problems (geocartography) by use of maps and computers, and those scholars whose research relates to the computer environment but who wish to apply their experience to "map data processing."

Much of this miscommunication was reflected in the 1979 NATO ASI in Maratea, Italy.

The first group, Spatial Analysts -- and notably geographers -- who may enjoy in-depth interaction with those who master the automated tools which make present cartographic spatial analysis so much more effective, seek, however, no assistance in methodologizing classifications of the data types with which they traditionally work, or the types of spatial problems which they encounter. But members of the other group, possibly because of the mathematical nature of their activity related to the computer, misjudge their best possible contributions to be indeed in the area of methodological thinking which they attempt to apply to geocartography and spatial analysis, which are foreign territories to them, conceptually.

It seems that a rich cartographic methodology related to a wealth of geographic theory are not sufficiently known or acknowledged. Consequently, construction of the tower by the two groups together, is handicapped by the lack of a meaningful communication as to what analytical mapping or map data processing are all about. It seems unclear what are the real areas of desired contribution of each group in order to advance automated map data processing.

It is the major purpose of this paper to discuss these questions from a methodological point of view. It will touch upon the very definition of cartography, which seems in itself to be increasingly blurring. It will evaluate the contribution of the computer to automated cartography. It will touch upon major data problems which form a bottleneck in automated geo-cartographic analysis. And finally, it will tie these aspects to problems related to automated spatial information systems.

THE CARTOGRAPHIC DOMAIN

Cartography is a general name for an extremely wide area of activity today. So wide and versatile is this area that workers, all identifying themselves as cartographers, may work on matters lacking totally an intellectual common denominator. The total cartographic domain may be subclassified in several different ways. One, seemingly helpful in our case, is presented in Table 1.

Activities within Cartography per se would be of a great variety of types. Some would be mechanical or technical, relating to photography, drafting, scribing, reproduction, and the like. Other activities may relate to visual or psychological aspects as symbolism, coloring, shape perception, and the like. Their common denominator is derived from the fact that contributions made in these activity areas result in general improvements of maps, as such, regardless of their theme or application purpose.

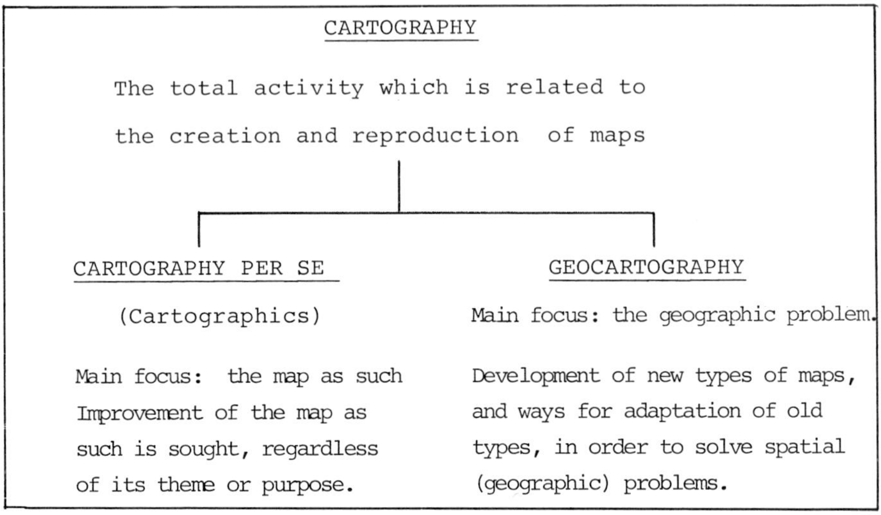

Table 1

Geocartography derives its core from geographic theory. In many ways has cartography been geography's conceptual laboratory. For many basic spatial concepts have been identified, classified, tested, elaborated on, understood in depth, and theoretically defined within the realm of cartography - maps and mapping. Or as Bunge puts it, maps "have been the logical framework upon which geographers have constructed geographic theory. They have been to geography what mathematics has been to some other disciplines." (1).

To mention a few examples, classical geocartographic literature provides in depth discussions on the concept of scale, levels of generalization, division methods of space, the concept of relative space, dimensions of data types,

---

[1]Bunge, William. "Theoretical Geography," C.W.K. Gleerup, Publishers, Lund, Sweden, (1962), p. 33.

classifications of spatial data types, class intervals, notions of shape and size, and a variety of quantitative analytical mapping approaches to fit data types and types of problems.

Spatial problems to which analytical mapping is applied are, of course, numerous, even in types. They are a great deal of what the geographical literature deals with. The following briefly mentioned examples may help demonstrate to non geographers the nature of the spatial problems which are conventionally solved by geographers, utilizing a variety of automated map data processing methods:

* Relations between parts of space (e.g. amount of seasonal labor migration between regions, or countries).
* Division of space into functional parts (political, administrative, by proximity, accessibility, etc.)
* Optimum location (planning for)
* Optimum spatial allocation (e.g. of services).
* Network topological analyses (e.g. for transportation planning).
* The influence of cartain spatial characteristics (e.g. distance) on... (e.g. students' choice of a university).
* The changing influence over space of the amount of precipitation, slope, and rock type, on the erosion and consequent stability of the coastal cliffs.
* Spatial attributes which influence the spatial distribution of (e.g.) crime in a given area.
* A Monte Carlo simulation of the spatial pattern and extent of (e.g.) Tel-Aviv in the year 2000.

Even these brief examples suffice to demonstrate why huge data banks, holding millions of bits of topographic D.T.M. portrayals of vast parts of earth, do not provide the data required to handle most spatial analyses which are relevant to the solution of these problems. Yet these are the real problems characterizing geographic research and planning. Relatively little of what is presented prevalently in map data processing meetings is related to all of this. This calls seemingly for a reassessment of the meaning of the computer to mapping and to spatial analysis by means of automated mapping.

CARTOGRAPHY + COMPUTERS = ?

Table II, provides a Classification Framework within which the Contribution of the Computer to Mapping may be Assessed Methodologically.

Research stage in which contribution is made	Level of contribution caused by automation			
	C*	B**	A***	
Observation (descriptive)	Dot maps Choropleth maps	Graduated symbol maps 3-D maps	Cartograms	Examples: Map Types
Analysis	Isoline maps Choropleth maps	3-D maps Dasymetric maps Projections	Centographic Portrayals Cartograms Homoplethic maps	
Theory (portrayal of theoretical stages or conditions of space).	Isoline maps	Projections	The above + other relative space portrayals; Monte Carlo simulation, gaming.	

Table 2

*C = Old map forms, automated - <u>potential</u> (of information content) <u>unchanged</u>

**B = Old map forms, automated - <u>increased potential.</u>

***A = New map forms, made possible due to automation - <u>new potentials.</u>

It is helpful to view the contribution of the computer to mapping in reference to the three classical stage of research recognized in the general methodology of scientific inquiry: observation, analysis, and theory.

In each of these stages maps may be, and are, utilized in geographic research. The "observation" stage is where data are compiled and are recorded spatially, or undergo initial, and basic, spatial arrangement prior to analysis.

The "analysis" stage is rather self explanatory. The "theoretical" stage is where generalized or abstracted portrayals of reality are created (models), or theorectical (e.g. anticipated, or hypothesized) portrayals are constructed. Obviously, continued developments in geographic theory, have been followed complementing developments in mapping methods.

Another facit of our methodological analysis concerns the level of the contribution to mapping achieved due to the application of the computer to the mapping process.

Three such levels, I suggest, may be observed. They are marked A, B, and C in Table 2.

In the first level stand old map forms which ahve been automated. They may have gained consequently in terms of execution speed, possibly in graphic aesthetics, probably in line-work precision, and the like. But the potential of their information content remains unchanged. That is, to use an example, Von Humbold's isotherms are not drawn today, some hundred and fifty years later, and with a computer, in a more meaningful way or in a way whereby the isotherm map (or generically, the Isarithmic = Isoline map) provides now more information, or an improved form of information.

In the second level stand, again, old and conventional forms of mapping; however, ones whose potential has been increased since they have been automated. For example, the graduated symbol map which was traditionally produced using several class intervals into which all data have been generalized, is now produced with the computer in a way whereby each symbol is drawn to its precise proportional size. Another example is provided by the Three Dimensional (3-D) maps which now, since being automated, are used in a great variety of simulation analyses - and not only because of the increased speed of their drawing.

In the third level stand new forms of mapping which came into being since, and due to, the employment of computer technology in mapping and in spatial analysis. For example, problems of relative space mapping have been solved, and the automated cartogram is one of their outcomes[2]. There are more examples, such as the Continuous Scanning method for Homoplethic mapping[3]. However, discussion of these examples is beyond the purpose of this paper.

An obvious conclusion is that the most significant future contributions of automated mapping should take place in level A - new map forms, automated, providing new potentials.

It is hoped, however, that developments and advances in automated mapping methods will concern rather equally all three research stages. For the "observation" stage whereby data are compiled or retrieved prior to analysis, has been rather neglected in the past. This is a major issue in what is termed in

[2] Tobler, W.R. "Computer Cartograms." Unpublished monograph, Ann Arbor, Michigan (1967).

[3] Degani, Avi. "Some Computer and Isodensitracer Applications in Geography." Journal of the Minnesota Academy of Science, 36 (1969-70)

the continued discussion "the data crisis".

THE DATA CRISIS

The bottle neck of geocartographic analysis is composed of data problems. Dieter Steiner has phrased it so adequately in the Maratea NATO ASI, "for the problems we'd like to solve we lack the data, and the data we have are good for nothing."

That is actually to say that so often we lack the data which may be turned into the information we wish to produce, and the data we have are very often available in forms which are non-transferrable into the necessary spatial information.

It may be well to emphasize first the essential difference between data and information. It is also an essential distinction relating to the difference between data banks and information systems, which is often unrecognized.

Data - are a raw state of compiled facts.

Information - are data which have been transformed (upped) into a (pre)defined level of meaning, in the context of need and purpose (application).

An example of a map portraying the spatial arrangement of a set of data is provided in Figure 1.

Shown is the distribution of crime, classified by major type, in the Hatiqva neighborhood, Israel's most notable slum area, in the southern edge of Tel-Aviv.

In order to acquire this set of data several severe barriers had to be overcome.

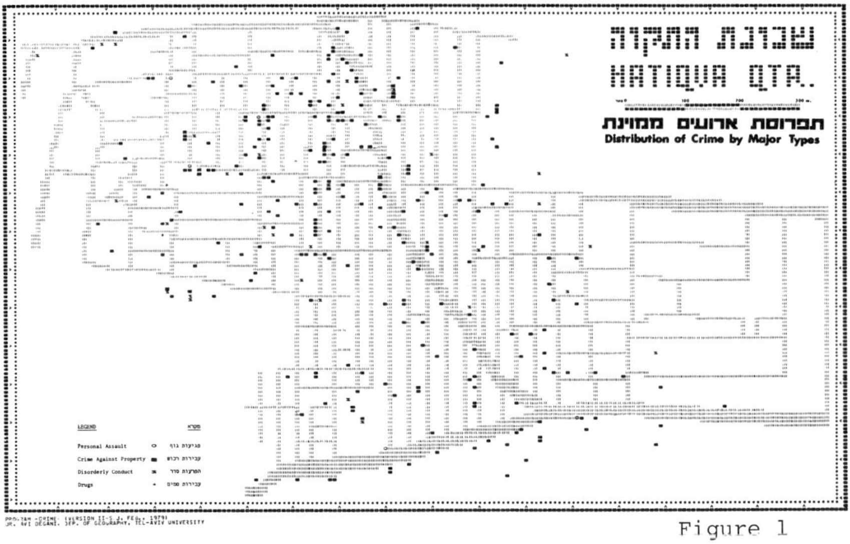

Figure 1

These barriers are typical of what is termed here the "data crisis", which is underlined by three major problems:

1. Most data compiled by most societies are compiled in area forms, while point data are obviously superior in terms of spatial analysis.
2. Most data are available in non-spatial forms.
3. Our analytical power accedes our compilation ability.

In reference to our particular case in point, the police in Israel (and probably everywhere) does not compile its crime reports but in areal form or at best by streets (under the name of the street). Namely, these are data on a nominal basis and non spatial. All reported crimes would have had to have coordinates assigned to them while registered in the police booking, to become potentially available for spatial manipulation and analysis.

Consequently, in order to transform these nominally referenced data into a spatial form, computer algorithms must be produced to perform such transformation. Many interesting problems are involved in the development of such algorithms. They may be interesting from a computer science point of view as well as from the mathematics relating to probability theory and spatial sampling. Yet there is hardly any significant work done along these lines. The creation of such algorithms, would have been a substantial step in the direction of closing the gap between our rather advanced analytical power (the wealth of analytical computer algorithms) and our poor compilation ability, hereby examplified.

In the particular case of the Hatiqva neighborhood crime was studied as part of a general project aimed at planning the rehabilitation of the neighborhood. The spatial information sought about crime was the locations and distribution of "crime zones."

According to an operational concept which guides the work of the police in this area, a "crime zone" is defined by N reported crime instances within an area enclosed by a radius of R meters, within a given period of time, P. This is a classical example where concept, need and purpose join to define the required information to be generated as based on the crime point data shown in Figure 2. Such data, as mentioned earlier, are not commonly available in a spatial form. But as a joint effort of the Tel-Aviv police and the slum rehabilitation reseach team, the police has compiled all reported crime data in a special form, including coordinates, for the length of the research period. It became then immediately possible to utilize such analytical models as the Continuous Scanning method in order to turn these data into the desired information (Figure 2).

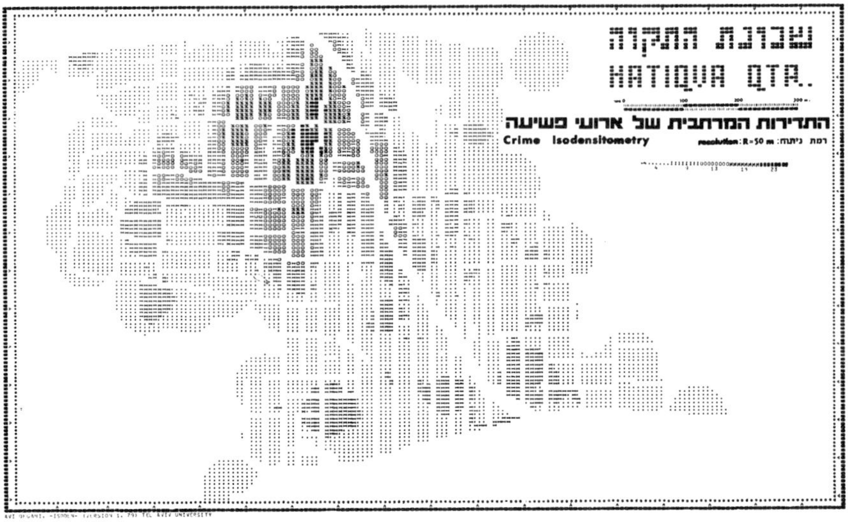

Figure 2

AUTOMATED SPATIAL INFORMATION SYSTEMS

All matters discussed thus far converge in the subject of automated spatial information systems. Let us define, first, this entity.

An automated spatial information system is:

* A dynamically updatable body of data (dynamic "data base," or "data bank"),
* interfaced with a body of models programmed for a computer transformation (computational + graphic + cartographic) of these data into spatial information,
* in order to satisfy specific needs
* of defined users
* within the framework of well defined concepts and technology.

Several major conceptual observations ought to be made concerning such systems:

a) Since the information system is based on a data base,

it must store data but must not store information.

b) An updating system must be built into the system, or it is not to be constructed at all. There are numerous examples of such data banks turning into "data graveyards."

c) The compiled ans stored data must be aimed, in their level of detail, to the predefined and expected needs of the system. Their accuracy must be within the good sense of need and purpose. Again, there are numerous examples where this logical rule is being rediculously violated, at high cost in all possible ways[4].

d) It is essential that alternative ways be evaluated to replace the prevailing approach of mass storage of enormous quantities of data. Two possible alternatives should be examined particularly. The first, the development of fast automated, data retrieval methods whereby instead of storing the information content of entire topographic sheets, for example, the necessary information alone will be compiled prior to use, and only from the relevant maps or their relevant sections. This should solve not only the problems of first time compilation, then storage, updating checks, and continued maintenance, but also will allow, prior to every use, compilation of the necessary data only, in the necessary resolution, which may, of course, vary from one application to

[4] See for example: Degani, Avi. "Spatial Systems and Automated Geocartographic Analysis," Proceedings of the 3rd Jerusalem Conference on Information Technology (1978).

another. Secondly, methods for the increased use of spatially sampled data should be evaluated to replace the prevalent trend of compiling total possible sets of data at such high costs. In a great many subjects the level of accuracy required in the projected analyses for which data are stored, may be readily accomplished with data values which will be created by interpolation methods, or the like.

e) It is important that every data set incorporated into a data base will be examined carefully as to the logical meaning of the expected information to be generated as based on it. Consider an example whereby "soil type" is stored for a projected use in ground stability checks related to building; However, "depth of top soil" is the data type rather needed. This is an excellent sample of critical, prevalent, mistakes that happen when the person assigned the top responsibility for the system is a master of the tool rather than a master of the subject matter for which the tool is being constructed.

Much of this experience has been gained in the methodological construction and planning of ISRAMAP - the Information System for Regional Automated Mapping Analysis and Planning, proposed for Israel.[5]

---

[5] Degani, Avi. An Automated Information System for Urban Multi Purpose Application in Israel (1977).

IN CONCLUSION

A host of problems has been discussed in this paper, all relating to geocartographic spatial analysis; however, relating to it in a variety of different levels. Problems of all levels converge at the spatial information system level, for it contains all the way from data to its analysis - in respect to both hardware and software matters. Future avenues of research, and professional communication between workers in various disciplines who take interest in map data processing, must be basically guided by a true understanding of what cartography is all about, what is spatial analysis, and where does the computer contribute to it, and on what level in each regard.

Data problems described call for much work which is essential in order to eliminate the troublesome bottle neck which stands in the way of much potential research, which may utilize sophisticated method of automated geocartographic analysis. It seems that presently too much energy is devoted to the compilation of physical data while social data, in their broadest sense, are almost totally ignored. This may be the doing of Landsat, and what can, and what cannot, be obtained from those images. I dare suggest that possibly much too much attention has been captured by what is available as sources of data (e.g. Landset images) on the expense of a search for ways to obtain the essential needed social data. thus far these data are almost unobtainable by use of automated methods. Several lines of research have been suggested concerning typical problems of automated spatial information systems.

THE TRANSFER OF SOFTWARE SYSTEMS FOR MAP DATA PROCESSING

Richard Baxter

Building Research Station
Department of the Environment
Garston, Herts, England

I. INTRODUCTION

This paper is set against a background of rapid growth and marked development in techniques for handling and analysing spatial data, a phase in which the role of computer technology has been highly significant. The initial impact of the computer in geography, cartography, planning and related disciplines was in statistical analysis and mathematical modelling, but in more recent years it has played an increasingly important role in the acquisition of spatial data, and in its storage, management and display in map form. There is now a substantial and distinctive literature dealing with computer-aided handling of locationally-referenced data, geographical information systems, and computer mapping.[1]

A consequence of these developments has been the proliferation during the last decade of computer software, and it may seem not unreasonable to assume that sufficient software is now readily available to satisfy almost any requirement relating to map data processing.[2] In principle, such a case can be convincingly made, since

ISBN 0-12-267180-5

solution methods for numerous analysis and processing needs are well known. In practice, however, there are serious problems for individuals or organizations wishing to import and implement a substantial computer-based capability for analysing and manipulating spatial data. There are several reasons for this, stemming both from the inherent nature of computer software and from the context of much of its development over the past ten years.

The general case for transfer is built upon the premise that software, data and personnel locked in upon a single computer manufacturer or installation inhibits the exchange of information, the replacement of equipment and competitive procurement. It can be argued that two types of software exist: those of general applicability and those of local concern. The former category could be candidates for exchange amongst users if logistical problems could be overcome. If exchange is to be effective, then certain standards are necessary. How might these be defined? These questions are now explored.

II. TRANSFER OF SOFTWARE SYSTEMS

The benefits of software exchange rest on six beliefs:[3]

i. it prevents a "reinvention of the wheel"

ii. everyone gains

iii. it is cheaper than redevelopment

iv. makes badly needed capabilities available to all

v. allows for a quick "technological fix"

vi. costs are dependent upon technical factors which are the major barriers to transfer.

The countervailing arguments are equally well rehearsed:

i. management and professionals say that in-house skills are enhanced by reinvention.

ii. job satisfaction amongst programmers suffers because implementation is less exciting than development.

iii. cost savings can occur but require particular conditions, the absence of which are only recognized with hindsight.

iv. not everyone really needs the capabilities.

v. the number of applications available and acceptable for local circumstances is small.

vi. in practice it is human frailties which are the major obstacles to effective transfer.

In spite of the reservations, a case can be made for software exchange in the field of map data processing. It rests on the rising costs of development because salaries and machine-time charges are inflating, problems are increasing in complexity, the volumes of data require greater processing efficiency, users have increasing expectations and compatibilitv between systems is a desired objective.[4] However, evidence suggests that software exchange is more successful when both donor and recipient have high levels of knowhow and expertise rather than in the traditional model where the exchange is a means of improving the recipient's knowledge and expertise or circumventing the need for local knowledge and expertise to carry out specific tasks.

Before discussing the details of software exchange it is as well to face up to the general problems. Computer software, designed with a specific purpose in mind, tends to incorporate features which derive from the nature of the data and original problem for which it is intended; these factors restrict its general usefulness. Similarly, restrictive characteristics may derive from the programming language in which the software is written and from the type of computer configuration on which it is first implemented. Each of these aspects introduces its particular dependencies, and taken together they generate restrictions which compound each other and result in software which is to a variable extent specific to the problem, data and machine environment from which it originates. Such software can be difficult to transfer and apply to other situations without considerable re-programming effort. In spite of these well known problems, generality and portability of much existing software for spatial applications have not rated at all highly in the criteria applied to its design and implementation.

Nor has the methodological context of the bulk of software development for spatial data encouraged its transfer to other situations and users. Generally speaking, the first priority has been to find computer solutions to the distinctive and immediate problems of spatial data handling and analysis without high regard for the generality and portability of these solutions. In consequence software development has in many respects proceeded in an uncoordinated, haphazard and diffuse manner in this field. This tendency has been compounded by the slow emergence of agreed and coherent methodological framework, reflected for example in the continued lack of concensus on matters as basic as terminology, data definition and data structuring. This has seriously detracted from the exchange and integration of different computer-based approaches and solutions to spatial data problems.

A common additional restriction stems from the fact that good software documentation is required for successful transfer; documentation not only of its implementation and usage, but, equally important, of the algorithms around which the software is built. Only rarely does more than minimal credit or satisfaction derive from writing good documentation, and so it is often done poorly or not at all. This is particularly the case for software written in an academic or research environment.

A problem which parallels and exacerbates these issues is the limited interaction between research into spatial problems and its application in spatially-oriented activities such as planning, transportation and environmental management. There are, for example, few effective means for channeling software research and development effort into practical applications, or to have applications significantly influence the course of research. Developments in software methodology on the one hand, and in applied methods of spatial data processing on the other, have tended to proceed independently and this has not encouraged the emergence of a coherent body of software recognised and accepted both by researchers and practitioners alike.

There has in consequence been considerable duplication of effort in the production of software for handling geographical data, while in spite of this investment there remains a low level of software portability and exchange, and a lack of accord on any generally accepted methodological standards for algorithm development and technology transfer.[5]

A. Nature of the Product

When discussing software transportability it is useful to distinguish between three types of application-oriented software: packages, programs and routines.[6] Whereas packages are designed to facilitate easy user interaction they are frequently cumbersome to move between different makes of computers. Routines can be made more easily transportable but require a degree of programming competence in their use. Programs sit somewhere inbetween in both respects.

Packages are designed to be general in their application and in consequence to cater for a wide user market. This generality imposes penalties when dealing with specific applications such as map data processing where even simple tasks with large data files can command considerable CPU and backing store resources. Generality apart, there are other reasons which inhibit the easy transfer of packages. The software is likely to be written in assembly code, possibly call upon library routines, handle all input and output, and interact with the operating system. Any transfer of such a package to a different computer is bound to be problematic in the first instance and will inevitably involve the specialist computer scientist in the process. Although the original documentation is likely to be good, there may be difficulties in implementing the package to perform strictly in the prescribed manner which may cause confusion to the peripatetic user.

Programs, on the other hand, are more specific in their application and in consequence have a smaller potential user group. They have usually been written by users and, through exchanges, made available to other users.[7] Transportability becomes a problem for a series of reasons. Firstly, there may be a difficulty in physically transferring the program between machines because of hardware incompatibilities or differences in input codes. This may pose a real problem if the physical transfer is left under the control of an inexperienced user. Secondly, there might be problems in compiling the program either because a suitable compiler is non-existent or because the source code is written in an extended (machine-dependent) version of a high level programming language. In the former case the program must be ignored and in the latter case the offending statements replaced by standard language equivalents. Thirdly, when the program is eventually compiled, it remains to test whether it runs. At this stage further problems may arise - demands on core space may be excessive (possibly involving overlays or paging), reference may be made to non-available subroutines which existed in some library, or use may be made of non-configured peripherals. Finally, even if a run is successful, the results may differ because of hardware characteristics (the parameters for real arithmetic being particularly significant). In spite of these obstructions, the capable programmer can usually succeed in getting the program to run if he is willing to expend sufficient effort. This effort is usually magnified by inadequate documentation but sometimes eased by being able to contact the program originator.

Routines are much more portable between different computers than are programs or packages; they can become virtually machine independent if care is taken to exclude installation-specific aspects and to follow a carefully defined set of programming conventions and language standards. The modular functional form of a routine library offers greater flexibility to the user of an applications program, who can select and combine routines in a variety of ways and can apply them to quite different data sets and problems. The routine approach by definition draws attention to the algorithm used to perform a specific task, which is significant from the point of view of a critical appreciation of methodology and use. The development and programming of routines enables software research and implementation to proceed in definable and manageable units, and to respond with flexibility and reasonable speed to new requirements. A very significant argument in favour of the routine library approach, particularly with regard to its feasibility, is the existence of several successful precedents in other numerical and computer-oriented disciplines.[8] The best of these prove attractive from the user's viewpoint and an effective method of using limited resources.

Whatever approach is adopted, the relationship between the software and the map data is important. Spatial data has distinctive properties and tends to be voluminous. The largeness arises for two reasons. Firstly, the area of coverage and information content demanded by users is extensive. It is common practice to use national coverage and on occasions it is necessary to seek global coverage. The second reason relates to resolution. For example, the number of coordinates necessary to define the boundary of a polygon can be very large if the polygon is highly convoluted and a precise representation is sought. If, at the same time, the number of polygons is large (e.g., land-use parcels in the country) then the total file can approach gigantic proportions. These factors enhance the case for efficient software and its dissemination.

B. Standards

Whether approaching software interchange through packages, programs or routines it is necessary to make some decision about programming languages. The choice is between a high-level language or a low-level language written for an abstract machine.[9] Portability is improved by adopting a stable high level language but their specifications tend to be incomplete, ambiguous or both.

To achieve maximum software portability in the field of map data processing, the widely implemented FORTRAN language (ANSI Standard) can be used for applications software. Not all machines and compilers interpret the ANSI Standard in the same way; so a number of restrictions are desirable to provide compatibility of software across a range of commonly used computers. Because many implementations tend to deviate from the standard, which is the "floor" for compiler design, it is important not to use any such extensions or undefined constructions and to check code using a verifier (e.g., PFORT).[10] An alternative is to use a good structured pre-processor (e.g., FORTRESS) or language translator (e.g., PROTRAN) which will produce standard code. A review of these standards will be necessary as the new 1977 ANSI Standard FORTRAN is released and implemented, but the major impact of these changes will be to relax certain restrictions and to make further language facilities available.

Whatever the choice of programming language, interchange is facilitated by prescribing conventions for code layout and documentation. It is a well proven rule of thumb amongst users of borrowed or stolen software that the usefulness is in strict proportion to the number of lines of legible documentation multiplied by the number of standard programming features in the coding. Some suggested standards for FORTRAN coding have been published.[11] There are a large number of qualities that may be expected of a good computer program. Amongst these are robustness, accuracy, predictability, clarity, maintainability, efficiency, modifiability and ease of use. If in the process of attaining these qualities longer, slower or less elegant software is produced, then this should be considered a necessary evil to be traded against the overall benefit derived from the potential transportability of the software.

Documentation must be as meaningful to the user as the software is to the computer. It serves three principal functions. Firstly, it aids the successful implementation of software from one computer installation to another. Good documentation is necessary if software is to be completely portable, and is absolutely essential where programming standards have been neglected. Secondly it enables the recipient of software to judge whether he has implemented it successfully. This means that documentation must include test input and associated output, against which particular software performance in a new environment may be evaluated. Thirdly, it enables a user to modify software so that it will conform more closely to his own particular requirements. Documentation is necessary to make enhancements, debug as necessary and interface with any input and output. To satisfy these needs in an orderly fashion, different types of documentation are required. A useful guideline is to provide a user manual, a reference manual, technical description, system implementation and a publicity brochure[12] to supplement code documentation and test data sets.

The testing of software for implementation purposes requires the provision of test data and specimen results which should be supplied with the software to any receiving installation. The supply of appropriate test data sets is important; they provide a significant means of validating the mounting and implementation of the software on any computer. Sets of test data should be developed which are easy to transfer between machines as card image character sets; are representative of different data notations and include differently scaled measures; test extreme cases; are for meaningful variables derived from existing data sets, relating if possible to one or a few study areas for purposes of illustration and comparison; and are accompanied by illustrations and descriptive statistical information, in addition to the calling programs and results. Documentation is particularly important for data sets. The procedural history should be given explaining the methodology, the coding nomenclature (and any deviations), the precision, the resolution, the reliability, the coverage and the currency.

C. Alternative Procedures for Exchange

The simplest approach to software interchange is by providing the facilities whereby the suppliers and customers can be put in touch with each other. This involves the compilation of software inventories from the inputs of suppliers and the dispatch of these up-to-date inventories to potential customers. The largest such inventory is that produced by the IGU commission on geographical information and remote systems. It is distributed on microfiche by Buffalo State University. Tarrant's publication on computers in Environment Sciences provides a somewhat similar service. More specialised arrangements exist through the advertisement of available programs in Environment and Planning[13] and Computers and Geosciences and the informal exchange in special interest groups.

The next step involves the marriage between a supplier and customer and the physical exchange of software. When no third party is concerned, the success of reimplementation depends upon the goodwill of the supplier and the degree of mismatch of the respective computer environments. For example, Buffalo and Cambridge Geography Departments exchange software with little difficulty because of their mutual interests and possession of similar mini-computers. The transfer of the GIMMS system[13] between Edinburgh and University College London and the Transport and Road Research Laboratory has succeeded because of personal commitments and in spite of hardware incompatibilities. A wider diffusion of computer program listings (rather than in machine-readable form) is achieved through journals[14], departmental published papers (e.g., Computer Applications from Kansas State Geological Survey and the University of Nottingham, Moray House College of Education, CATMOG, and working papers) and books.[15] All these sources make software available subject to varying degrees of implementation difficulties which it is left to the recipient to resolve.

A more rational approach sees the establishment of some focus through which software is exchanged. At its simplest it would merely acquire software and documentation and pass them on to the customer. The supplier benefits by moving the administrative chores elsewhere but the recipient might suffer if problems occur during reimplementation and the software originator is not readily at hand. This deficiency can be rectified by the centre developing and maintaining some expertise in all the programs it handles and making this available on request. The Geography Program Exchange originating in Michigan and now housed at Buffalo is an archetype. Whilst distribution is widespread (in spite of unit charges for each program) there is a degree of failures in re-implementation for the reasons cited. The Harvard Laboratory for Computer Graphics and Spatial Analysis fulfils a similar function.

The GPE has spawned a European offspring in Zurich but the exchange traffic appears to be only localised and there is no knowledge of any foreign involvement. In Britain there is the international clearing house for segment oriented referencing systems (SORSA) stationed at the Transport and Road Research Laboratory. Although it has received a handful of large programs it has passed on very few programs over the years.

The centre that distributes software can be more ambitious and attempt to rewrite software according to established standards. Such activities usually extend to a deliberate selection policy, the preparation of standards and guidelines (cf algorithms in Environment and Planning), the creation of test data sets, and the provision of comprehensive documentation. The efforts involved in these tasks, especially for interactive programs to run on both mainframes and minis, are huge. Apart from fundamental problems of software standardization and resulting rewriting of code, there are questions of hardware compatibility, compiler consistency and tape code formats that tax the whole exercise.

A fundamental question concerns the physical transfer of software. In most of the situations cited, the most frequent medium for the transfer of software is probably the non-machine-readable listing. Although listings are necessary back-up documents, there are occasions (e.g., for large files) when it is preferable to transfer software on machine-readable media. Each of the three commonest transfer media (punched cards, paper tape and magnetic tape) have clear advantages and disadvantages associated with their use. Punched cards are robust in transit and easily read by standard input peripherals. Decks of interpreted cards can be read and edited manually. Against these advantages there are disadvantages: they are costly to purchase and mail, codes vary between machines and cards often contain a large proportion of unused space. Paper tape has the advantage of being cheap to buy and with minimal unused space is cheap and compact to mail. There is also a widely used standard. However, paper tape is easily torn and manual editing is fussy; reader units often tend to be inaccurate. Magnetic tape has a high storage capacity with low bulk and at relatively low cost. The disadvantage is the bewildering variety of standards and recording conventions. Other devices, such as cassettes or discs, do not enjoy sufficient universality at present to make them suitable media for exchange. Whatever medium is used, care has to be exercised about codes (e.g., use of tabs) and layout.

The point has already been made that map data tends to be voluminous and for this reason exchange is more effective by using a computer readable medium. Costs, reliability and compatibility dictate that this should be a magnetic tape. Decisions need to be made about the file and physical structure of the data on the tape. As far as file structure is concerned, it is preferable to adopt sequential, fixed length files in character format using standard codes. The physical structure should use standard labels for exchange between the same machines (otherwise no labels) on 9 track tape recorded at 800 or 1600 bpi using EBCDIC or ASCII codes. Even if these conventions are adhered to, there is no guarantee of a painless transfer.

In the future networks and distributive processing will allow software (and data sets) to be readily transferred between machines or accessed remotely. Apart from facilitating actual exchange, networks make it possible to share loads and resources between computers. Load sharing allows work to be done on machines with spare capacity or hardware better suited to the task whilst maintaining savings over communication costs. Resource sharing concerns the availability of, inter alia, software on specific installations which can be accessed remotely or transferred to the local environment. The transition will pose many problems of a logistical (e.g., legal) and technical nature.[16]

D. Costing and Ownership

Much software exchange to date has been on a friendly give and take basis. When charges are involved the sums are nominal. The question of ownership has rarely arisen.

However, when applications become successful or the financial climate deteriorates, the picture may change rapidly. Rather than buy proven software on a commercial basis, the evidence is that most agencies prefer to redevelop equivalent software in-house, thus reducing efforts to exchange products with a common usage.

The experience of the Social Science Research Council Survey Archive in Britain over pricing is salutary. Even token subscriptions reduced the demand and it was assessed that the user community would not readily pay for software. This has significance insofar that any software exchange centre cannot hope to be self-financing. It is also likely to have more uncertain repercussions. Whereas the individual supplier might be willing to service a small market on a materials repayment basis, he is likely to change his attitude when the demand increases. The increase in demand will not only require more handling time, but, because the level of sophistication amongst the customers will drop, he will be pressured for advice. The expansion of the user base may place further demands upon updating, modification and documentation, and the situation may rapidly get out of the supplier's control. This, or the mere change of employment, might result in the software being abandoned and not maintained before there is sufficient understanding amongst the user community to continue its use.

The ownership issue could be controversial. Parent organizations (e.g., universities), funding bodies or commercial establishments might make claims on the ownership of software written by their staff. They would presumably only seek to do this if they anticipated financial gain through the use or sale of the software or expected costs through the obligation to maintain software. This could influence the whole practicalities and costings of software exchange. The situation varies between countries; the United States, for example, places government-funded software in the public domain, whilst in Britain it is marketed through, amongst other agencies, the National Research and Development Corporation.

A not unrelated issue relates to the protection of software. At the moment the patent and copyright exist to protect the originator although the legal position is extremely complicated. Whereas patents have been applied, they are difficult to enforce and likely to be curtailed. Copyright is embodied in the expression of novel ideas. In the software context, the copyright is reserved to the program listing and not the background algorithm. This means that translation of programs infringes copyright whereas reprogramming does not. This has some importance for the exchange of software because of the ownership rights that the supplier might wish to preserve (for financial or other reasons). Again the situation varies between countries and awaits international agreement and clarification.

III. CONCLUSIONS

This paper has attempted to raise some of the issues concerned with the exchange of software systems. The point has been made that because needs (i.e., applications) are rarely identical there is a limit to the potential for software exchange. However, the underlying theme is that exchange is inhibited by human factors rather than technical. Whilst there is room for improvement in standards and documentation, the greater need is for a more enlightened attitude by management and professional towards the benefits of exchange.

There is little ambiguity about the products to be exchanged (software) but doubts remain as to the best means of effecting the exchange. Bilateral arrangements seem to work best and centralized clearing houses cannot be profitable without government subsidy. The question of costs is problematic given that there is a high price elasticity of demand. The question of legal ownership raise problems affecting the free exchange of products. All in all, the issue of exchange is more contentious than might at first sight appear.

ACKNOWLEDGMENTS

This paper draws extensively upon two reports prepared for the Social Science Research Council in Britain. The first of these was produced by a working party, of which the author was a member, which examined selected problems of computer software provision in research and practical activities dealing with spatial data. The second was commissioned from the author and was to evaluate the need for a software and data exchange centre to service the geography community. The views expressed are personal and do not necessarily represent the views of the author's Department.

FOOTNOTES AND REFERENCES

1. Overviews of the state of the art in these areas can be found in:
   Baxter, R. S., 1976, Computer and Statistical Techniques for Planners (London, Methuen).
   Tomlinson, R. F., Calkins, H. W., Marble D. F., 1976, Computer Handling of Geographical Information Systems (Paris, Unesco Press).
   Peucker, T. N., 1972, Computer Cartography (AAG Commission on College Geography, Resource Paper Series, 17).
2. See, for example, the extensive Software Index produced by the IGU Commission on Geographical Data Sensing and Processing as part of its work on information systems for state land use planning. It is available from the Department of Geography, State University of New York at Buffalo.
3. Kraemer, K., 1976, Transferring R & D Technologies and Lessons Learned in the United States, in Information Technology and Urban Governance (Ed Wellar, B.), Ministry of State for Urban Affairs, Canada.
4. Witiuk, S., 1976, in Proceedings of Workshop on Current Issues in Geographic Data Processing (Ed. Taylor, F.), Carleton University, Ottawa, Canada.
5. A measure of the significance of software portability issues, and a review of many of them, can be obtained from:
   Brown, P. J. (Ed.), 1976, Software Portability (Cambridge, Cambridge University Press).
   Muxworthy, D. T., 1976, A Review of Program Portability and Fortran Conventions, Technical Paper Series, 1,

European Computer Program Institute, Ispra, Italy.

6. The relative merits of alternative approaches are discussed in:
Nelder, J. A. 1974, Statistical Packages: Introductory Remarks,
Janacek, G., 1974, Some Observations on Design and Accuracy of the BIOMEDICAL Computer Programs Package,
and
Muxworthy, D. T., 1974, A Review of Some Statistical Packages, all of which appear in an appraisal of statistical packages in the Institute of Mathematics and Its Applications, 8, 165-174.

7. Best known perhaps are the Geography Program Exchange (now coordinated by the IGU Commission on Geographical Data Sensing and Processing) and the Laboratory for Computer Graphics and Spatial Analysis, Harvard University. Various other sources and reports publish software in printed listing form, and many departments of geography have program libraries.

8. For a review of alternative software information sources, including program libraries, see:
Langenheder, W., 1976, The present state in Europe of software access and information transfer in the field of research and teaching, Proceedings, First ECSIR Workshop, 14-22.
University computer users are probably most familiar with the NAG (Numerical Algorithms Group) Library, one of the most successful in this field. The organization, history and activities of the NAG project are discussed in:

Numerical Algorithms Group, 1975, NAG Annual Report (Oxford, NAG).

Ford, B., Hague, S. J., 1974, The organization of numerical algorithms libraries, in Software for Numerical Mathematics (Ed., Evans, D. J.) (London, Academic Press), 357-372.

Specific issues relating to software portability in the NAG context are discussed in:

Ford, B., 1976, The envolving NAG approach to software portability, in Software Portability (Ed Brown, P. J.), (Cambridge, Cambridge University Press), 249-267. Another example, in the computer graphics field, is GINO (Graphical Input and Output) originally developed in the Computer Laboratory at Cambridge University and subsequently implemented and distributed by the Computer Aided Design Centre, Cambridge. A description is given in:

Woodsford, P. A., 1971, The design and implementation of the GINO 3D graphics software package, Software:

9. Tanenbaun, A. S., Klint, P., Bohn, W. 1978, Guidelines for software portability, Software - Practice and Experience, 8, 681-698.

10. Ryder, B. G., 1974, The PFORT Verifier, Software - Practice and Experience, 4, 359-378.

11. 1977, Notes for authors on the preparation and description of computer algorithms, Environment and Planning A, 9, and Working Party on Statistical Computing, 1975, Construction and Description of Algorithms, Applied Statistics, 24, 366-373.

12. Jaro, M., 1978, Standards, definitions and interfaces: conditions for technology transfer, SORSA Workshop.

13. Waugh, T. C., 1973, An Introduction to a Geographic Information Manipulation and Mapping System, Research and Development Notes - No. 9, Edinburgh Regional Computing Centre.

14. An example of such a routine is presented in: Williams, I. N. and Baxter, R. S., 1977, Three Point Rational Function Interpolation for Calibrating Gravity Models, Environment and Planning A, 9, 215-211.

15. See, for example: Baker, L., 1974, A Selection of Geographical Computer Programs, Geographical Papers, 6, London School of Economics; Baxter, R. S., 1976, Computer and Statistical Techniques for Planners (London, Methuen); Davis, J. C., 1973, Statistics and Data Analysis in Geology (New York, Wiley); Dawson, J., Unwin, D., 1976, Computer Programming for Geographers (London, David and Charles); Harbaugh, J. W., Bonham-Carter, G., 1970, Computer Simulation in Geology (New York, Wiley); MacDougall, E. B., 1976, Computer Programming for Spatial Problems (London, Edward Arnold); Marble, D. F., 1967, Some Computer Programs for Geographic Research, Geography Department, Northwestern University; Mather, P. M., 1976, Computers in Geography (Oxford, Blackwell); Merriam, D. (Ed.), 1966, Computer Contributions, Kansas State Geological Survey; Rushton, G., et al, (Eds.) 1973, Computer Programs for Location - Allocation Problems, University of Iowa, Geography Monograph 6; Till, R., et al, 1971, A Collection of Computer Programs in BASIC for Use in Geology and Geophysics, University of Reading, Geology Report 5.

16. See Baxter, R. S., 1979, Computers in Planning - Past, Present and Future, Proceedings of International Congress on the Applications of Computers in Architecture, Building Design and Urban Planning, Berlin.

17. American National Standards Institute, USA Standard FORTRAN X 3.9 - 1966, United States of America Standards Institute, New York. Highways Engineering Computer Branch, 1979, HECB Programming Instruction Manual, Volume III, Language Standards: FORTRAN, Department of Transport. (The National Computer Center Standard FORTRAN Programming Manual forms the core of this publication, with additions by HECB).

PATTERN RECOGNITION PROBLEMS IN THE CLASSIFICATION OF MULTI-IMAGES

Silvano Di Zenzo

IBM-Italia

Pattern recognition techniques are used in digital picture processing mainly for classification of the whole picture into one of a specified set of classes. In the case of multispectral imagery PR techniques are also used for partitioning the image into pairwise disjoint subsets, based on the spectral signature of each pixel. Some of the problems which arise in the latter respect are reviewed, with emphasis on iterative techniques. A well known algorithm for component labelling is adopted for obtaining a partition of a multi-image of interest in scene analysis.

I. PATTERN CLASSIFICATION OF ASSEMBLIES

A major problem in the application of pattern recognition techniques to digital picture processing is that of estimating a probability density function f(x), where $x=(x_1,..,x_n)$, best fitting an n-dimensional histogram of observed data.

In the last decade, efforts have been concentrated on the estimation of f(x) as a mixture of normal densities; techniques for obtaining maximum-likelihood estimates of the parameters involved have been studied by many authors (see, for example, 3-5-13).

There exists a "natural iterative procedure" suggested by likelihood equations obtained on equating to zero the partial

ISBN 0-12-267180-5

derivatives, taken with respect to the signature parameters, of a certain log-likelihood function.

Unfortunately, such natural procedure may converge to a solution which is not a maximum-likelihood estimate.

The convergence of the natural procedure has been investigated by W.A. Coberly and B.C. Peters in the case in which the a priori probabilities are the unknown parameters (the more interesting case for supervised classification of multi-images). These authors have shown that the scheme is locally convergent, i.e. converges to the maximum-likelihood estimates of such probabilities for starting values sufficiently near such estimates (2).

In a series of papers (9-12) a modified iterative procedure has been presented and discussed in detail; the modified procedure is shown to converge locally to the maximum-likelihood estimates of the signature parameters α_i, μ_i, Σ_i (α_i weighting coefficient, μ_i vector of means, Σ_i covariance matrix of the ith normal distribution in the mixture).

These theoretical results seem to be quite relevant. However an open problem remains: how near an initial signature estimate must lie to a maximum-likelihood estimate in order for the iterates to converge to a maximum-likelihood estimate.

In a recent paper (4) the present author has given sufficient conditions for global convergence of the unmodified iterative procedure to the maximum-likelihood estimates of the α_i's, when these are the only unknown parameters.

A brief account will be given in the sequel. It is the author's opinion that such conditions cannot be appreciably weakened. They are extremely heavy, and it seems unrealistic that they may be satisfied in any practical situation. On the other hand, most of the benefits brought by any iterative procedure are given by the first few iterates (there is experimental evidence for this fact). It seems that a limited and supervised use of the iterative approach is advisable, confined to the case in which α_i's are the only unknown parameters.

In the author's opinion the advantages of the iterative

scheme are mainly methodological. It offers, in a sense, a reference scheme of pattern classification for those cases in which a large assembly of individual objects has to be classified (as is the case for pictures, considered as assemblies of picture elements). In what follows, the recognition problem will be reformulated under this viewpoint. The natural iterative procedure will be derived by a consistency argument rather than from the log-likelihood function (this is possible when the α_i's are the only unknown parameters). A brief review of the subject of supervised classification of multi-images will also be given.

A. The Recognition Problem Restated

Let us recall here the general formulation of the problem of pattern recognition in terms slightly more abstract than usual. A recognizer O can be schematized as a set of measuring (or sensing) devices $T_1,..,T_n$, together with some processing capability. When the investigation is pressed to some depth, the difference among measuring and sensing devices shows to be immaterial.

Each T_k can measure a certain well defined magnitude τ_k, for example the intensity of light in a given bandwidth.

By no means are the τ_k confined to be physical magnitudes, for example they could be questions in a questionnaire $(\tau_1,..,\tau_n)$: the discriminant analysis problem would require peculiar computational tools in this case, but the situation is in principle the same.

By an observation of an individual object B made by O, we mean the measurement, made by O, of the values that $\tau_1,..,\tau_n$ take at B. It is intended that B is a value-bearing individual for the compound variable $\tau=(\tau_1,..,\tau_n)$; i.e., $\tau_1,..,\tau_n$ are among those variables which characterize the structure of B as a complex object.

Let us agree that B is an observable object for our recognizer O whenever B is a value-bearing individual for $\tau=(\tau_1,..,\tau_n)$.

Now suppose that B is also a value-bearing individual for

some other variable ξ unobservable by O. ξ is just another one of the variables which characterize the structure of B: it only happens not to be observable, i.e. measurable, by O.

ξ cannot be directly measured by O, but, if there is some kind of dependence between ξ and τ and, moreover, O is allowed to observe (i.e. measure τ on) a certain number of individuals whose value of ξ is known, then O can develop some procedure for evaluating the probability that ξ=y for a given individual, if its observed value of τ is x.

Here x$\in$R and y$\in$S, where R and S are the spaces where τ and ξ take their values, respectively. R is the so called pattern, or feature, space.

Example 1. τ_1 is the weight and τ_2 the height of adult human beings; ξ is the sex. If ξ=1 for males and ξ=2 for females, then S={1,2}. In this case, to be referred to as the "sex case", O is unable to distinguish sex directly; it can only test hypotheses about sex on the basis of the observed values of the weight and height. Of course O has to be trained to recognize sex from weight and height.

Example 2. Observable individuals: again adult human beings; $\tau_1,..,\tau_n$ the questions of MYNNESOTA psychological test (n=500). Only two answers are allowed for each question, thus R=$\{0,1\}^n$, a metric space which calls for special computational tools. ξ might be the following 3 level categorization of the ratio between height and weight: ξ=1 for lean, ξ=2 for medium-size, ξ=3 for fat people. MYNNESOTA has never been used for this job, however, if even a slight but peculiar difference exists between the average psychological profiles of lean and fat people, then one may expect that, after an extensive training, the recognizer will be able to make good inferences about the degree of fatness of people on the sole basis of their response to the test.

Example 3. Observable individuals: 80 meter square elements of the earth surface; $\tau_1,..,\tau_4$ the reflectivities in the international bands 4,5,6 and 7 of the electromagnetic spectrum (the bands used by the Multispectral Scanner System on Landsat 1 and 2). The recognizer is a remote sensing device on board an aircraft plus some image processing facility. The unobservable variable ξ might be a variable whose possible values label various types of land cover: e.g. $\xi=1$ water, $\xi=2$ dirty water, $\xi=3$ bare soil ,.., $\xi=m$ other. The problem is of course that of assigning each surface element to one of the classes. There are a number of practical as well as theoretical difficulties and some open questions.

The most striking feature of putting things this way in pattern recognition is that perception fits a questions-and-answers scheme in which questions are something which is built in the perceiving subject. In the first example the objects to be classified are inquired about their height and weight, in the second about a number of their private things, in the third about their optical behavior. In each case the tested object gives a response, i.e. a set of answers. We shall be mainly interested in the optical response of surface elements.

Let us briefly remind how the recognition problem is approached by means of statistical discriminant analysis (see, for example, 5-6).

As a first step, the recognizer has to be trained, i.e. allowed to observe (measure, sense) $\tau=(\tau_1,..,\tau_n)$ on a (large) number of individuals whose value of ξ is known. Then it perhaps will be able to obtain m density functions $f_1(x),..,f_m(x)$, one for each class in which the individual objects have to be classified. $f_i(x)$ should be an estimate as good as possible of the conditional probability $f_i^\circ(x)=P(\tau=x|\xi=i)$ that an individual belonging to class i, that is, bearing the value $\xi=i$, bears the value $\tau=x=(x_1,..,x_n)$.

In the case of scene analysis, each $f_i(x)$ is obtained on the basis of a set of training fields for class i, that is, a sample of picture elements almost surely representing surface elements belonging to the ith class of land cover.

In most cases the estimation of the density functions is made under normality assumption, so that sample data is used for calculating a covariance matrix and a vector of means for each function.

When the training is over, the actual classification is

performed in a straightforward way: an individual whose observed value of τ is x is assigned to the class i whose density function f_i has the largest value at x. If the a-priori probabilities of encountering individuals of each class are equal and the costs of misclassifying the ith class as the jth are independent of i and j, then the decision procedure stated above actually minimizes the expected cost of misclassification.

The procedure above has a drawback: each individual object is classified independently from the others. For picture processing this is a major drawback. Considered as assemblies of picture elements, pictures are far from being structureless sets: there is an adjacency relation defined among pixels, carrying with it such features as edges and lines, arcs and curves, end and branch points, and such properties as textural, topological and shape properties.

When we classify "pixel by pixel", on the mere basis of the spectral signature of each individual pixel, i.e. its observed value of τ, we simply ignore all of that. The type, if not the amount, of the information that is lost this way can be realized quite easily. The probability that a given pixel belongs to the class "water" should be reinforced if its neighbors seem to belong to "water". If a pixel lies on a line (the standard reference for terminology is 10), the probability of its being homogeneous to other pixels on the line should be reinforced; conversely, any evidence in favour of this homogeneity should, in a sense, reinforce the hypothesis that the line is a line.

It seems however that a satisfactory supervised classification procedure taking in full account both the spectral signatures and the adjacencies is still to be produced.

In the next subsection a fact which may at first seem surprising will be illustrated: the classifying procedure stated above is not correct for an assembly of objects, even when the assembly is considered to be a simple, structureless set.

B. Classification of Sets of Individual Objects

Let $A=\{A_1,..,A_r\}$ be a set of value-bearing objects for both ξ and τ; $\xi=1,2,..,m$, while τ is assumed to take its values in R^n, that is, the space of n-tuples of real numbers.

With reference to Example 3 (the case of Landsat imagery), A may be thought as a portion of the earth surface and $A_1,..,A_r$ as its decomposition into surface elements. To each A_k there corresponds a pixel in an image: disregarding adjacency, only the spectral signature remains, i.e. a vector $X_k=(X_{k1},..,X_{kn})$ of observed radiances in n spectral bands. X_k can be considered as an observation of the compound random variable $\tau=(\tau_1,..,\tau_n)$. Under this viewpoint $\{A_k\}$ is a sample obtained in without-replacement, hence non-bernoullian scheme. As it will be seen, the recognizer must guess, for each $i=1,..,m$, the fraction of pixels belonging to class i within the sample.

With reference to the "sex case" (Example 1), $A=\{A_1,..,A_r\}$ will represent a sample of human beings. By no means does A have to be a random sample. The recognizer doesn't know anything about A: it can only measure weight and height and, starting with some a priori estimates for the fractions of females and males in the sample, use its computing power in order to evaluate the probability that each A_k be male or female, the expected value of the number of males in the sample, and even the probability that this number be p, $p=0,1,..,r$. The value of $\tau=(\tau_1,\tau_2)$ observed at A_k will again be denoted X_k. In this case X_k is an ordered pair of real numbers (X_{k1}, X_{k2}).

Here, in order to perform the classification, the availability of the densities $f_i(x)$ is not sufficient: the recognizer should also manage to obtain an estimate p_i of the probability $P(\xi=i)$, for every $i=1,..,m$. Here $P(\xi=i)$ is simply the probability of encountering an individual of class i within the sample, i.e. the fraction of such individuals in the sample. The obvious condition

$$\sum_{i=1}^{m} p_i = 1 \qquad (1)$$

must be satisfied. The estimates p_i are used as weights when the actual classification is carried out: for j=1,..,r, A_j is assigned to class h if

$$p_i f_i(X_j) < p_h f_h(X_j) \qquad \forall i \neq h \tag{2}$$

This classification rule follows from Bayes theorem, which in the actual context can be written

$$P(\xi=i|\tau=x) = \frac{P(\tau=x|\xi=i)P(\xi=i)}{\sum_{k=1}^{m} P(\tau=x|\xi=k)P(\xi=k)} \tag{3}$$

This holds true for every $x \in R^n$ and i=1,..,m. The left hand side is the conditional probability that an object whose observed value of τ is x belongs to ith class.
None of the probabilities which appear in the right hand side of (3) is known: we can only obtain estimates $f_i(x)$ of $P(\tau=x|\xi=i)$ and p_i of $P(\xi=i)$, for i=1,..,m. On replacing these estimates for the corresponding "true" probability densities in (3), we are able to obtain an estimate p_{ij} for the probability $P(\xi=i|\tau=X_j)$ that jth object belongs to class i:

$$p_{ij} = \frac{f_i(X_j)p_i}{\sum_{k=1}^{m} f_k(X_j)p_k} \tag{4}$$

For fixed j, p_{ij} attains its maximum value when $f_i(X_j)$ is maximum: hence (2).

We shall now derive the natural iterative procedure by a straightforward consistency argument. As the sample $\{A_k\}$ is completely non-bernoullian, i.e. coincides with the universe it has been drawn from, in order to be compatible among themselves the estimates p_i must satisfy the following condition

$$p_i = \frac{1}{r} \sum_{j=i}^{r} p_{ij} \tag{5}$$

Indeed p_i has to be equal to the average of the individual probabilities p_{ij}. (5) coincides with the first of the likelihood equations. Likelihood equations are obtained along a different line of reasoning. $\{X_k\}_{k=1}^{r}$ is considered as an independent sample of observations drawn from a linear convex combination

$$f(x) = \sum_{i=1}^{m} p_i f_i(x) \tag{6}$$

of probability density functions $f_i(x)$ of given analytical form. Then, one manages to obtain a maximum-likelihood estimate of the parameters contained in (6), i.e. values of such parameters which maximize the log-likelihood function

$$L = \sum_{k=1}^{r} \log f(X_k) \tag{7}$$

On equating to zero the partial derivatives of L with respect to parameters and performing the calculations, one obtains certain equations involving parameters: these are the maximum-likelihood equations. Their form quite naturally suggests an iterative scheme for their solution: this is known as the "natural iterative procedure". Whichever be the analytical form choosen for the densities in (6), (5) is always one of the maximum-likelihood equations. So, whichevery way we derive it, (5) is free from assumptions on the analytical form of the mixtured densities.

A few comments seem convenient. The two approaches to (5) are equivalent, despite the first treats $\{X_k\}$ as a completely non-bernoullian sample, while the second as an independent sample. However the first seems more appropriate for pattern classification of assemblies.

Apart from greater simplicity, it stresses that, when we are classifying an assembly of objects, that assembly becomes a universe by itself.

It stresses that $P(\xi=i)$ is not an abstract probability that $\xi=i$, but the probability that this event holds true within the assembly.

With reference to Example 1, $P(\xi=2)$ is not the probability of encountering a woman in the world, but in the sample $A=\{A_k\}$; it is 0 for a regiment of grenadiers, 1 for the Salvation Army of Great Britain.

In Example 3, $P(\xi=3)$ is proportional to the extent of bare soil in the particular scene, not on the planet.

All of this may sound quite trivial, but it does not seem that its implications have always been considered carefully: e.g., when partitioning the pattern space into regions corresponding to pattern types one should not forget to attach weights p_i to his discriminant functions.

For the sake of clarity, the natural iterative procedure based on (5) will be stated explicitly:

(a1) Select m a priori estimates $p_i^{(0)}$ of the probabilities $P(\xi=i)$,

(b1) compute first order estimates $p_{ij}^{(1)}$ of the probabilities $P(\xi=i|\tau=X_j)$ by replacing p_i's by $p_i^{(0)}$'s in (4).

(a2) Obtain first order estimates $p_i^{(1)}$ of the probabilities $P(\xi=i)$ by (5),

(b2) compute second order estimates $p_{ij}^{(2)}$ by replacing p_i's by $p_i^{(1)}$'s in (4).

(a3), (b3), etc.

For each i=1,..,m, the sequence

$$p_i^{(0)}, p_i^{(1)}, p_i^{(2)}, .. \tag{8}$$

is convergent. Indeed, it is non-negative, bounded by 1, and monotone (as can be verified easily). However the problem is not with the convergence, but with the limit.

In the next subsection a brief account will be given of the numerical problems connected with the limit values of sequences (8). This subsection will be concluded just mentioning another possible approach to the problem of obtaining maximum-likelihood estimates p_i. We shall suggest the underlaying idea by the following simple example.

Consider the "questionnaire case" of Example 2 and let

$n=1$. Let $\xi=1$ for males and $\xi=2$ for females. The questionnaire (only 1 question) is submitted to each individual in a sample $A=\{A_k\}_{k=1}^{r}$ and r_0, r_1 answers 0 and 1 are obtained. The problem is that of estimating the number of males and females in the sample.

For each $i=1,2$ and $k=0,1$ the training of the recognizer had provided an estimate $f_i(k)$ of the probability that an individual of sex i gives the answer k. Then the maximum-likelihood estimates p_1 and p_2 are obtained from the following consistency conditions

$$rp_1f_1(k) + rp_2f_2(k) = r_k \quad , \; k=1,2 \tag{9}$$

Notice that, for $f_1(0)=f_2(0)$, which implies $f_1(1)=f_2(1)$, any discrimination is impossible (case of extremely bad separation, see the following). (9) can be generalized to more complex situations in a relatively easy way. However again numerical problems are encountered. A method to overcome these is being experimented at the IBM Image Processing Center in Rome.

C. Convergence Results

As said before, a brief summary of results about the global convergence of the sequences (8) to the maximum-likelihood estimates q_i will be given. That will perhaps provide some insight into the reasons why (8) is not likely to converge to q_i. The following abbreviations will be adopted

$$f_i^\circ(x) = P(\tau=x|\xi=i), \; p_i^\circ = P(\xi=i), \; f^\circ(x) = P(\tau=x)$$

It can be shown that sufficient conditions for global almost sure convergence of (8) to q_i for each $i=1,..,m$ are the following:

(a) for each i, the estimated density $f_i(x)$ coincides with the true one, i.e. $f_i(x)=f_i^\circ(x)$, and then $f^\circ(x)=\Sigma_i f_i(x)p_i^\circ$,

(b) for each i, the number of objects of type i within the sample is large (in the sense of the law of large numbers),

(c) the densities $f_i(x)$ are pairwise different almost everywhere (i.e. no two of them coincide in a whole open interval).

Notice that (a) implies $q_i=p_i^\circ$.

It is understood that the initial values $p_i^{(0)}$ are choosen in the open interval (0,1) and their sum is 1.

A brief discussion seems convenient. Preliminarly, notice that τ can be assumed as having a discrete, or even finite, set of possible values. In plain words, (b) means the following: let r_i be the true number of individuals of class i in the sample $(r_1+\cdots+r_m=r)$, then the number of individuals of class i whose observed value of τ is x can be assumed to be $r_i f_i(x)$. Of course (b) implies this to be true only in the limit.

If the given conditions are not satisfied, (8) may jump over q_i on passing from an iterate to the next (more precisely, the given conditions are sufficient for excluding this). If this jump takes place, as the sequences are monotone, (8) will keep on departing from q_i.

It can be shown that, under (a)-(c), the m sequences (8) become stationary if they take exactly the m-tuple of values $q_1,..,q_m$ at some iterate.

Under (a)-(c), a "jump over" may not take place.

D. A Few Theoretical Remarks

We shall pay some attention in order to keep different concepts well separated. As seen, there are problems in getting (8) to converge to q_i: a distinct problem is that of the difference between q_i and $P(\xi=i)$.

For i=1,..,m, the maximum-likelihood estimate q_i may be different from $P(\xi=i)$ for more than one reason.

First, q_i may differ from $P(\xi=i)$ due to the fact that $f_i(x)$'s are not accurate estimates of the corresponding "true" densities $f_i^\circ(x)$'s.

This in turn may depend on a number of things: the training fields have not been choosen properly in the supervised classification of an image, the $f_i(x)$'s are constrained to an analytical form which does not suit reality, the training samples were too small.

Now suppose that problems in getting (8) to converge to q_i, as well as differences between $f_i(x)$'s and $f_i^\circ(x)$'s have been removed. Can we say that (8) converges to $P(\xi=i)$? The answer is no.

On the contrary, we would come to the conclusion that in any case we would be able to find out, as limits of sequences, the fractions of pixels of each type i in the image. That would be absurd: consider for example the case of two types of soil $i \neq j$ completely indistinguishable for the given remote sensing device (henceforth we confine to classification of multi-images, however all applies to classification of assemblies).

To attain a positive answer a further requirement must be met: $f_i(x)$'s have to be pairwise different almost everywhere. All of that would provide an almost sure convergence of sequence (8) to $P(\xi=i)$ for each i.

This brings under attention the "degree of separation" among $f_i(x)$'s taken two by two. Some insight may be obtained through information-theoretic considerations.

According to a relatively recent viewpoint in information theory (8), let $H(\xi)$ be the uncertainty on the random variable ξ after the remotely sensed data is made available for a given scene

$$H(\xi) = -\sum_{i=1}^{m} p_i^\circ \lg_2 p_i^\circ \tag{10}$$

Let $H(\xi|\tau)$ be the conditional uncertainty on ξ if the observed value of τ is known, and analogously for $H(\tau|\xi)$. We shall need the latter

$$\begin{aligned} H(\tau|\xi) &= \sum_{i=1}^{m} p_i^\circ H(\tau|\xi=i) = \\ &= -\sum_{i=1}^{m} p_i^\circ \int_R f_i^\circ(x) \lg_2 f_i^\circ(x)\,dx \end{aligned} \tag{11}$$

Here $dx=dx_1 dx_2 .. dx_n$. $H(\tau)$ is analogously defined as the uncertainty on the value of τ with reference to a given multispectral image:

$$H(\tau) = -\int_R f^\circ(x)\lg_2 f^\circ(x)dx \tag{12}$$

Besides ξ and τ, the compound variable $(\xi,\tau)=(\xi,\tau_1,..,\tau_n)$ may be considered. For (ξ,τ) also a residual uncertainty may be specified for the image under investigation, and the fundamental relation holds

$$H(\xi,\tau) = H(\xi|\tau) + H(\tau) = H(\tau|\xi) + H(\xi) \tag{13}$$

Taking into account (10),(11),(12), we obtain then

$$H(\xi)-H(\xi|\tau) = \sum_{i=1}^{m} p_i^\circ \int_R f_i^\circ(x)\lg_2(f_i^\circ(x)/f^\circ(x)) \tag{14}$$

The left-end side represents the average decrease in the uncertainty about ξ (i.e. the increase of the informatioon about the value of ξ) due to the knowledge of τ. More precisely: the average decrease in the uncertainty about the value that ξ takes at a generic surface element due to the knowledge of the optical response of the element, i.e. its observed value of τ.

A brief discussion of (14) may be useful. Remembering that

$$f^\circ(x) = \sum_{i=1}^{m} p_i^\circ f_i^\circ(x) \tag{15}$$

we see that the greatest lower bound of (14) is 0 while the least upper bound is (10). GLB is reached when all the f_i° coincide almost everywhere (extremely bad separation): in this case ξ is stochastically independent from τ and the knowledge of τ for a surface element says nothing about its value of ξ.

LUB is reached when, for each i,j, $f_i^\circ(x)$ is almost everywhere 0 in every interval where $f_j^\circ(x)$ is not such (in plain words: when the probability of misclassifying i as j is 0 for every i,j=1,..m): in this case ξ is functionally dependent on τ. In this same case $H(\xi|\tau)=0$, hence the knowledge of the value taken by τ at a given surface element destroys all the residual uncertainty about its value of ξ (and permits an almost sure inference of such value).

II. UNSUPERVISED CLASSIFICATION OF IMAGES

In the unsupervised classification of images no training fields are provided and one manages to find out a partition of the image, considered as a set of pixels, based on some nearness criterion. Clustering algorithms are used (1-12). ISODATA has played a prominent role so far (see (6) for a recent account).

If a notion of distance is introduced among pixels, it is then possible to define the internal deviation of any subset of the image as the sum of the distances between the elements of the subset taken two by two. Then also the total internal deviation of the whole image is a well defined magnitude. Clustering algorithms aim at partitioning the image into pairwise disjoint subsets in such a way that the sum of the internal deviations of the separate subsets is as small as possible with respect to the total internal deviation of the image.

There is an underlaying belief that pixels near with respect to the given distance will represent surface elements similar under various aspects, e.g. land cover.

For better insight we must be more specific. Given two positive integers k,m, let K be the set of positive integers not greater than k, and M that of nonnegative integers less than 2^m. A digital multi-image is a map φ: $K \times K \rightarrow M^n$, where M^n is the cartesian power of M by itself n times.

A pixel in the image is an ordered pair (a,b), where $a \in K \times K$ and $b=\varphi(a)$; b, which is an n-tuple of m-bit numbers, is the spectral signature of the pixel. The whole image can be identified with the set I of its pixels.

Less formally, a digital multi-image is a matrix whose entries, called pixels, are n-tuples of m-bit integers; however the more abstract definition, which will be adopted here, allows for greater conciseness.

A distance in I may be defined in more than one way. Let, for example, d be the city block distance in $K^2 = K \times K$ and let

$$d((a,b),(a',b')) = d(a,a') \qquad (16)$$

for every pair of pixels (a,b) and (a',b') in I. Then (I,d) is a metric space.

Again, let ρ be the city block distance in M^n and let

$$\rho((a,b),(a',b')) = \rho(b,b') \qquad (17)$$

for every pair of pixels (a,b) and (a',b') in I. (I,ρ) turns out to be a pseudometric space ($\rho(u,v)=0$ does not imply u=v), but it does not change much.

Clustering algorithms are often applied to (I,ρ), where ρ may be replaced by some other distance in M^n. This is the case for ERMAN II, a general purpose image processing program licenced by IBM. All the comments at the end of I.A apply again.

The next subsection contains the description of a clustering algorithm acting on (I,d,ρ) which has been implemented at the IBM Image Processing Center in Rome.

A. A Clustering Algorithm for Digital Multi-images

We shall need the reflexive closure of the adjacency relation. Two elements in I are adjacent (or 4-adjacent) when their d-distance is 1, contiguous when it is not greater than 1.

In plain words, two pixels with same row (column) number and column (row) numbers differing by at most 1 will be said to be contiguous.

We shall say that two pixels are ε-similar if they are contiguous and their ρ-distance is not greater than some positive number ε to be specified by the user. ε-similarity turns out to be a reflexive and symmetric relation in I, hence its transitive closure, which we denote by R, is an equivalence relation.

This means that, given $x,y \in I$, xRy if and only if a sequence $x_1,..,x_n$ of pixels can be found in I such that $x=x_1$, $x_n=y$ and x_i is ε-similar to x_{i+1} for i=1,..,n-1.

The elements of the quotient set I/R, i.e. the equivalence classes of the equivalence relation R, will be called the macro-elements of the picture, or, briefly, macro-pixels.

The next step is to form clusters of macro-pixels. Again,

a notion of distance must be defined among macro-pixels. Let us agree that two macro-pixels are near when the d-distance between them is not greater than some positive number D specified by the user.

Rigorously, the d-distance between two macro-pixels A,B is the smallest of the numbers d(x,y) for $x \in A$, $y \in B$. To speed up computations, A,B may be replaced by their respective minimum circumscribed rectangles with edges parallel to the lines and columns of the image.

We can now introduce a reflexive and symmetric similarity relation among macro-pixels. We shall say that two macro-pixels are δ-similar if they are near and their ρ-distance is not greater than δ (a positive number specified by the user).

Again, the ρ-distance between two macro-pixels A,B is, by definition, the smallest of the numbers $\rho(x,y)$ for $x \in A$, $y \in B$: any intercluster distance in M^n may be used instead, e.g. the L1 distance between centroids.

Let S be the transitive closure of δ-similarity: S is an equivalence relation in I/R, that is, among macro-pixels. For each $F \in (I/R)/S$, we form the set-theoretic union of those macro-pixels which belong to F. The sets of pixels so obtained, which we call classes, are disjoint and their union is I: they form a partition of I. This partition of the image under investigation is the output of the algorithm.

An efficient algorithm for providing such partition is quite similar, in its general architecture, to the algorithm for component labeling by tracking runs of 1's described in subsection 9.1.3 (10).

ACKNOWLEDGMENTS

I am indebted to Professors B. de Finetti, L. Guerriero and A. Grasselli for scrutiny and suggestions.
Thanks are due to Drs. P. DI Chio, C. Jacob, F. Orti Daras for helpful criticisms and discussions.

My warm thanks to Professors H. Freeman and G. Pieroni

who gave me the opportunity of lecturing on this subject to an audience of experts. Finally, I am grateful to Mrs. Stefania Rossi for her elegant coding of the algorithm in II.A.

REFERENCES

1. Anderberg, M.R. (1973). "Cluster Analysis for Applications" Academic Press, New York.
2. Coberly, W.A., and Peters, B.C. (1976). "The numerical evaluation of the maximum-likelihood estimate of mixture proportions", Univ. of Houston, Report 50.
3. Day, N.E. (1969). "Estimating the components of a mixture of normal distributions", Biometrika 56, p. 463-474.
4. Di Zenzo, S. (1979). "Extimating areal extents in supervised classification of multi-images", Proc. AICA Conference 1979, Vol. 3.
5. Duda, R.O. and Hart, P.E. (1973). "Pattern Classification and Scene Analysis", Wiley, New York.
6. Enslein, K., Ralston, A. and Wilf, H. (1977). "Statistical Methods for Digital Computers", Wiley, New York.
7. Grasselli, A. (1969). "Automatic Interpretation and Classification of Images". Academic Press, New York.
8. Kolmogorov, A.N. (1968). "Logical Basis for Information and Probability Theory". IEEE Trans. Inf. Th., Vol. II-14, n. 5.
9. Peters, B.C., and Walker, H.F. (1976). "An Iterative Procedure for obtaining ML estimates of the parameters for a mixture of normal distributions". Univ. of Houston, Report 51, NAS-9-12777.
10. Rosenfeld, A., and Kak, A.C. (1976). "Digital Picture Processing". Academic Press, New York.
11. Steinhausen, D. and Langer, K. (1977). "Clusteranalyse". Walter de Gruyter, Berlin, New York.
12. Walker, H.F. (1976). "The numerical evaluation of ML estimates of the parameters for a mixture of normal distributions". Univ. of Houston, Report 54, NAS-9-15000.
13. Wolfe, J.H. (1970). "Pattern clustering by multivariate mixture analysis". Multivariate Behavioral Research, 5.

A COMPARATIVE TEXTURE CLASSIFICATION EXPERIMENT

Joan S. Weszka[1]

IBM Corporation
Austin, Texas

This paper contains a comparative study of techniques for terrain classification based on texture. Features are derived from a texture's power spectrum, cooccurrence matrices and difference histograms. Results indicate that power spectrum features lead to overall lower classification rates than the cooccurrence-based or difference histogram features.

I. INTRODUCTION

A central problem in image analysis is the automatic classification of image parts based on texture. One of the earliest applications of texture analysis to be investigated was terrain classification. Haralick (1) contains a recent and extensive review of research on texture classification and analysis.

A variety of approaches to measuring textural properties in images have been developed over the years. This paper will discuss three approaches to texture analysis, and the results of applying them to a database of terrain samples of three geological types. The first approach computes features

[1]This research was partially supported by funds derived from the Air Force Office of Scientific Research under Contract F44620-72C-0062, while the author was at the Computer Vision Laboratory, University of Maryland, College Park, Maryland.

ISBN 0-12-267180-5

from the texture's Fourier power spectrum. The second is based on statistics derived from the grey level cooccurrence matrix i.e., on joint probability densities of pairs of grey levels, and the third approach is based on statistics derived from the probability density of grey level differences.

Feature sets were chosen so as to make them comparable with respect to their orientation and size sensitivity. Classification results are compared for all pairs of features from each feature set. These results were originally reported in (2).

The source of the database used in this study was a LANDSAT image of Eastern Kentucky (frame E1354-15424, band 6) and is shown in Figure 1. Dr. John R. Everett, a geologist at the Earth Satellite Corporation, identified and demarcated three regions on the image. These regions correspond to the following geological types:

1) Mississipian limestone and shale (A),
2) Lower Pennsylvanian shale (B), and
3) Pennsylvanian sandstone and shale (C).

From each of these three regions 60 samples were selected. Each of these samples was 64 by 64 pixels. The grey scale of each sample was linearly scaled to achieve a range of 0-63. The samples were then subjected to a "histogram flattening" transformation so that all grey levels occurred with equal probability. Figures 2-4 contains the 180 normalized windows. The motivation for applying the histogram transformation was to remove the effects of unequal brightness and contrast in the original LANDSAT image. These effects could have dominated the computed feature values. A discussion of histogram flattening can be found in Rosenfeld and Kak (3).

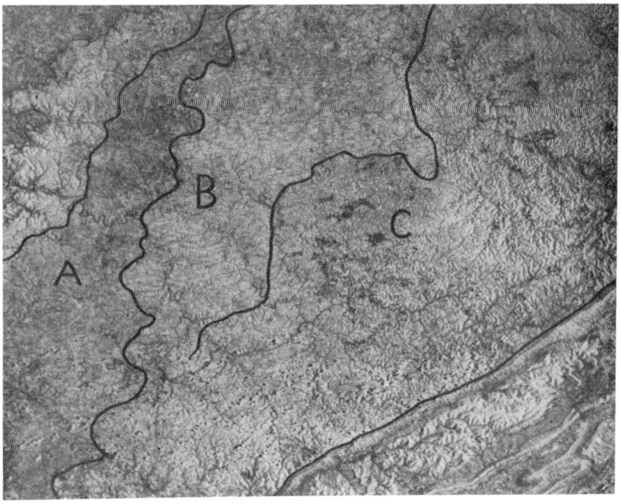

FIGURE 1. LANDSAT image used in study, with regions outlined. A-Mississipian limestone and shale. B-Lower Pennsylvanian shale. C-Pennsylvanian sandstone and shale.

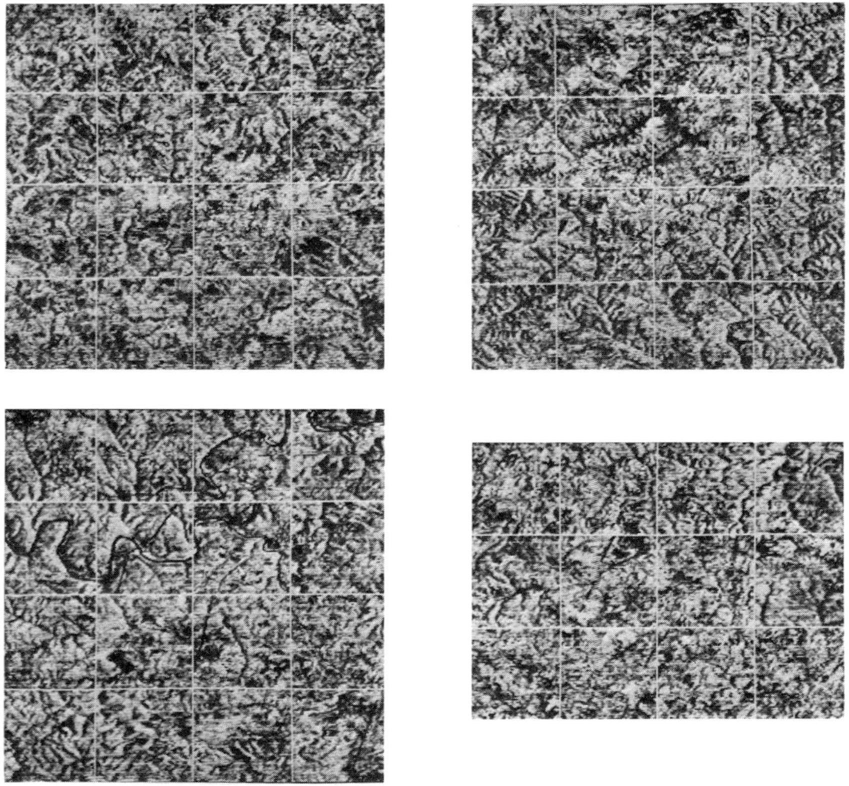

FIGURE 2. Windows of terrain type A (Mississipian limestone and shale).

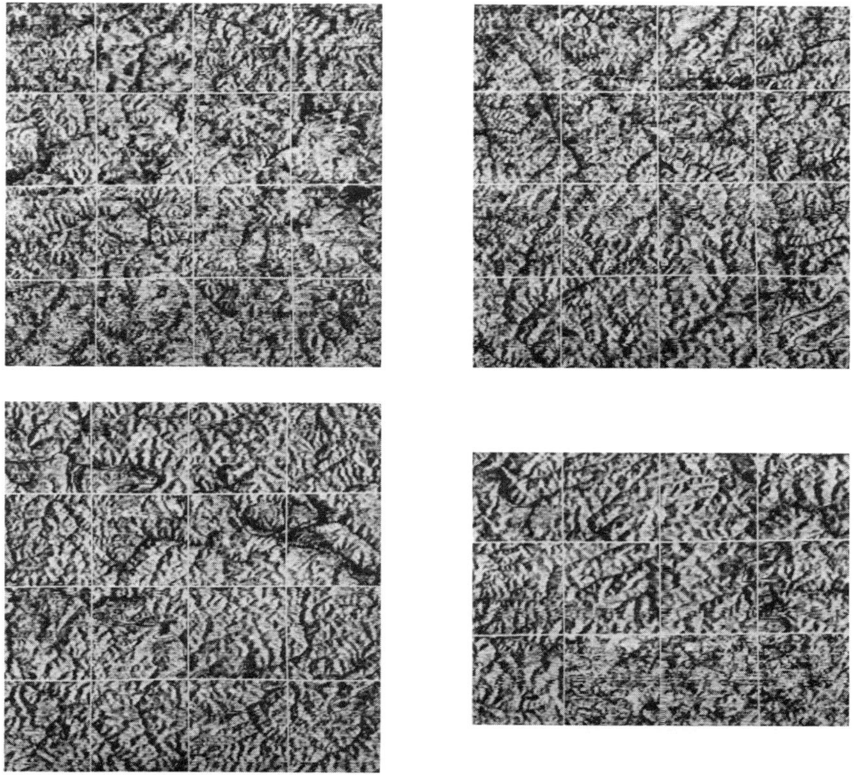

FIGURE 3. Windows of terrain type B (Lower Pennsylvanian shale).

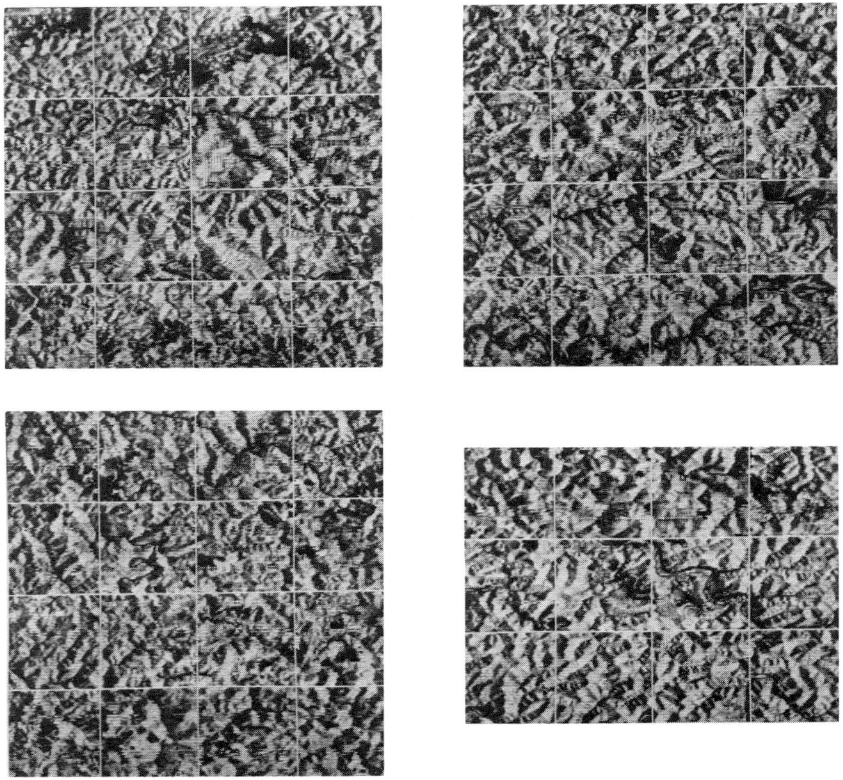

FIGURE 4. Windows of terrain type C (Pennsylvanian sandstone and shale).

The remainder of this paper is organized as follows: Section II defines the texture features that were used. Section III presents the results of the classification study. Finally, Section IV contains a brief summary.

II. TEXTURE FEATURES

This section describes the three classes of features that were computed on the database samples, and the particular feature sets used within each feature class. An attempt was made to equalize the feature sets with respect to their sensitivity to texture direction and coarseness. This was done so that the comparison of the classification results would be more meaningful.

The first class of features used was based on the Fourier power spectrum. For an n by n digital picture, f, the discrete Fourier transform of f is defined as:

$$F(u,v) = \frac{1}{n^2} \sum_{i,j=0}^{n-1} f(i,j) e^{-2\pi k(iu+jv)} \qquad (1)$$

where $0 \leq u \leq n-1$ and $0 \leq v \leq n-1$. This transform treats the input picture as though it were periodic. If, in fact, it is not, then the transform is affected by "edge effects" which are discontinuities that are created when the picture is wrapped so that its opposite edges are adjacent. These discontinuities introduce spurious horizontal and vertical directionality, which manifests itself in the power spectrum as high values along the horizontal and vertical axes.

The standard Fourier texture features are averages computed in ring-shaped and wedge-shaped regions of the power spectrum. For the discrete transform, these features are of the form:

$$\psi_{r_1 r_2} = \sum_{\substack{r_1^2 \le u^2+v^2 \le r_2^2 \\ 0 \le u,v \le n-1}} |F(u,v)|^2 \qquad (2)$$

where r_1 and r_2, the inner and outer radii, define a ringed-shaped region, and:

$$\psi_{\theta_1 \theta_2} = \sum_{\substack{\theta_1 \le \tan^{-1}(v/u) < \theta_2 \\ 0 < u,v \le n-1}} |F(u,v)|^2 \qquad (3)$$

where θ_1 and θ_2 define a wedge-shaped region.

The Fourier features used in the present study were averages taken in regions defined by the intersections of rings with wedges. The following values of r_1, r_2, θ_1, and θ_2 define the 16 regions used, i.e., the regions derived by intersecting the four rings with the four wedges:

	r_1	r_2		θ_1	θ_2
ring 1:	2	4	wedge 1:	-22.5°	22.5°
ring 2:	4	8	wedge 2:	22.5°	67.5°
ring 3:	8	16	wedge 3:	67.5°	112.5°
ring 4:	16	31	wedge 4:	112.5°	157.5°

Note that the wedges start at -22.5° rather than at 0° so that any spurious high values along the horizontal or vertical axes will lie within two wedges (1 and 3) rather than being split across all four wedges.

The second set of features used was based on the grey level cooccurrence matrix. For any two-dimensional vector $\underline{d}=(dx,dy)$ we can compute the relative frequency with which pairs of grey levels occur at separation $\underline{d}$ in a picture f. These relative frequencies can be represented in an n by n array, a, where n is the number of grey levels in f. The array is computed by first translating the vector $\underline{d}$ over f and incrementing a(j,k) whenever j and k are the grey levels at the end points of $\underline{d}$. For example, for $\underline{d}=(1,0)$, and the following 3 by 3 picture:

0	1	2
3	3	3
2	2	1

the unnormalized cooccurrence matrix, a, is:

	0	1	2	3
0	0	1	0	0
1	0	0	1	0
2	0	1	1	0
3	0	0	0	2

The entry in row i and column j is the number of times that grey level i occurs horizontally adjacent to grey level j.

The matrix can be normalized by division by the sum of all the elements in the unnormalized matrix. The resulting normalized cooccurrence matrix is denoted A.

A variety of features can be computed from the cooccurrence matrix in order to measure texture (4). In our study, the "contrast" feature, defined as follows, was used:

$$\text{CONTRAST} = \Sigma(i-j)^2 A(i,j) \tag{4}$$

The contrast feature is the moment of inertia around the main diagonal of the matrix. In our study the following vectors, d, with displacements, dis, were used:

d	dis
(1,0), (0,1)	1
(1,1), (1,-1)	$\sqrt{2}$
(2,0), (0,2)	2
(2,2), (2,-2)	$2\sqrt{2}$
(4,0), (0,4)	4
(3,3), (3,-3)	$3\sqrt{2}$
(8,0), (0,8)	8
(6,6), (6,-6)	$6\sqrt{2}$

Note that these vectors are in the horizontal, vertical and two diagonal directions, the same directions used for the

Fourier features. Similarly, the magnitudes of the vectors correspond to the spatial frequency ranges used in computing the Fourier features.

The third set of textural features used was based on grey level difference histograms. For any vector $\underline{d}$=(dx,dy) we can compute a difference histogram vector, h, where h(i) is the number of times that i is the absolute difference in grey level between pairs of points separated by $\underline{d}$. For example, if $\underline{d}$=(1,0), then h for the following image:

```
0 1 2
3 3 3
2 2 1
```

is:

i	h(i)
0	3
1	3
2	0
3	0

The difference histogram can be normalized by dividing h by the sum of its elements.

The difference histogram described above is an example of a local property histogram. A local property is an image property which depends only on the grey levels in small neighborhoods of a picture. Statistics computed from local property histograms include mean, variance, skewness and kurtosis. In our study, the mean of the normalized difference histogram, H, was used. This statistic is computed as follows:

$$\text{MEAN} = \Sigma i * H(i) \quad (5)$$

The values of $\underline{d}$ used to compute difference histograms were the same as those used to compute the cooccurrence matrices, i.e., 16 vectors consisting of four different magnitudes in each of four directions (horizontal, vertical and two diagonals).

III. COMPARATIVE CLASSIFICATION STUDY

The three feature sets described in Section II were designed to give equal treatment to size (or spatial frequency) and direction by using, in each feature set, four sizes (or spatial frequency bands) and four directions. Each feature set was used to classify the 180 samples.

The classification scheme used the Fischer linear discriminant (5). The set of features measured on a picture is treated as a feature vector $\underline{Z}$. Suppose we are given a pair of classes, C_i and C_j, and a set of feature vectors $\underline{Z}_{i1}$, $\underline{Z}_{i2}$, ..., $\underline{Z}_{ik_1}$, and $\underline{Z}_{j1}$, ..., $\underline{Z}_{jk_2}$, obtained from the pictures belonging to C_i and C_j respectively. Let the sample means and the sample covariances be $\underline{u}_i$, $\underline{u}_j$, and S_i, S_j, respectively. A linear discriminant direction, $\underline{t}$, is obtained such that when we project the feature vectors on this direction, $(u_i-u_j)^2/(s^2_i+s^2_j)$ is maximized, where u_i and u_j are the mean values of the projected samples from classes C_i and C_j, and s^2_i and s^2_j are their variances. It can easily be shown that the optimal linear direction is given by:

$$\underline{t} = (S_i + S_j)^{-1}(\underline{u}_i - \underline{u}_j) \tag{6}$$

To decide the class to which a sample, $\underline{Z}$, belongs, we compute $\underline{t}^T\underline{Z}$ and compare it against $(u_i s_j+u_j s_i)/(s_i+s_j)$. When we have more than two classes, say $C_1,\ldots,C_m$, we can use a voting scheme to classify a given measurement $\underline{Z}$. For each pair of classes, C_i, C_j, we project $\underline{Z}$ on the appropriate line and classify it as described above. This gives us m(m-1)/2 different classifications of $\underline{Z}$. Finally, we assign $\underline{Z}$ to the class receiving the most votes.

For each feature set, all of the sixteen features were used in pairs as input to the Fischer-based classifier. The number of samples classified correctly by the best feature pairs are shown for each feature set in Table 1. The best single feature classified 136 samples correctly (about 75%), compared to 167 samples (about 93%) for the best pair of features.

Comparison of the results in Table 1 shows that the Fourier features performed the poorest, whereas the other two feature sets performed about equally well. Since the MEAN feature computed on the difference histogram is more efficient to implement than the CONTRAST feature computed from the cooccurrence matrix, the MEAN feature would be the preferred choice.

The relatively poor performance of the Fourier features may be explained as follows:

1) The Fourier features were computed using the discrete Fourier transform, which treats the picture as though it were periodic, even if it is not. Thus, the Fourier features may have been affected by the spurious values introduced by discontinuities at the edges of the picture.

2) The textures of the terrain database may be more accurately modeled in the space domain (e.g., as random fields with specified autocorrelations) rather than in the Fourier domain as sums of sinusoids.

TABLE 1. Best Performing Feature Pairs

Feature Set	Feature Pair	Number Correct
Fourier Power Spectrum	1) Ring (16,31) Wedge 1 and Ring (16,31) Wedge 4	158
	2) Ring (16,31) Wedge 1 and Ring (16,31) Wedge 3	153
	3) Ring (16,31) Wedge 2 and Ring (8,16) Wedge 3	151
Cooccurrence Matrices	1) $\underline{d}$ = (1,1) and $\underline{d}$ = (2,0)	166
	2) $\underline{d}$ = (0,1) and $\underline{d}$ = (1,-1)	166
	3) $\underline{d}$ = (1,0) and $\underline{d}$ = (1,1)	163
Difference Histograms	1) $\underline{d}$ = (1,1) and $\underline{d}$ = (2,0)	167
	2) $\underline{d}$ = (1,0) and $\underline{d}$ = (0,1)	165
	3) $\underline{d}$ = (1,0) and $\underline{d}$ = (1,1)	164

3) Haralick (1) points out that for a texture which can be described as having been generated by a Markov process, the values of the autocorrelation can be computed from the grey level transition probabilities for nearby points. Thus, for such textures, the autocorrelation function contains less information than the grey level transition probabilities. Perhaps the textures investigated in this study can be modeled by such Markov processes. No attempt, however, has been made to test this assumption.

IV. SUMMARY

We have presented the results of a comparative classification study of image texture using features derived from the texture's power spectrum, cooccurrence matrices, and difference histograms. The results favor the features computed from difference histograms and cooccurrence matrices over those derived from the power spectrum, although further experiments are required before the results reported here can be used to reliably guide the design of an image texture classification system.

REFERENCES

1. Haralick, R., "Statistical and Structural Approaches to Texture," Proceedings of the IEEE, 67, 5, 1979, pp.786-804.
2. Weszka, J., Dyer, C., and Rosenfeld, A., "A Comparative Study of Texture Measures for Terrain Classification," IEEE Trans. Systems, Man and Cybernetics, 6, 4, 1976, pp.269-285.
3. Rosenfeld, A. and Kak, A., Digital Picture Processing, Academic Press, New York, 1976.
4. Haralick, R., Shanmugan, B., and Dinstein, I., "Texture Features for Image Classification," IEEE Trans. Systems, Man and Cybernetics, 3, 1973, pp.610-622.
5. Duda, R., and Hart, P., Pattern Recognition and Scene Analysis, Wiley, New York, 1976.

SEGMENTATION TECHNIQUES AND PARALLEL COMPUTATION FOR IMAGE PROCESSING

S. Levialdi

Laboratorio di Cibernetica, CNR
Arco Felice, Naples, Italy.

I. INTRODUCTION

We are being confronted by an ever growing amount of numerical data in all spheres of human activity and this trend does not seem to be changing, for the time being. In practice, numbers, printouts, report sheets, are coming out from electromechanical devices at rates which are hard to follow and the real question to ask when observing this phenomenon is whether all this is useful (in a broad sense) to improve man's living conditions rather than how shall we handle all this information in the best possible way. But the subject of this School is not concerned with the first question but, perhaps, with the second, although I strongly believe that different are the answers to the choice of the data processing strategy according to the evaluation one performs of the data to process.

The particular information which we will be dealing with comes from maps, photographs or any two dimensional data structure which represents physical properties of a real surface (earth's surface is an example). Photographs of the land taken from airplanes or images sensed from satellites in one, or many spectral bands, constitute important data sources since they are able to provide vast amounts of information (in bits) which must be processed automatically or semiautomatically by means of interactive systems.

As we shall see, some of the techniques which may be used for processing these images have grown from the experience gained when dealing with other images and, in fact, the whole activity of image processing is based on the assumption that most of the problems that have to be solved are general in nature, that algorithms may be improved if some modelling is done of the human operator and that if some knowledge of the specific nature of the images to be dealt with is embedded in the processing strategy the results will be improved. Examples of all this will follow. In conclusion every task in image processing may certainly have its own peculiarities but there is a core of general purpose techniques which may be adapted to every single problem, if not, new techniques must be developed using the experience gained until now. A printed evidence of this is the vast literature on image processing which counts to over 600 papers yearly, two scientific journals, (1,2) a special IEEE Transactions on this subject (3) and a number of international meetings which are held every year, where applications to biomedicine, remote sensing, optical character recognition, electron microscopy, industrial inspection, signature validation, etc. as well as mathematical tools for processing, classifying, coding and transmitting images are presented.

ISBN 0-12-267180-5

II. MAP IMAGES

Let us now briefly list the main disciplines which may be included in the remote sensing area. Refer to table I. in (4).

The list contains: geology, agriculture, forestry, colonization, environmental surveillance, meteorology, hydrology, oceanography, geodesy, cartography, topography, military surveillance and reconnaissance. Satellites like Landsat, Meteosat, etc. as well as aerial photography from specially planned flights are two of the main sources of this imagery; the devices used for collecting the data are highly sophisticated and will not be described here. Let us only mention that different spectral sensors are required for multispectral scanning, only daylight is allowed for photography which has also to take place at definite hours of the day; furthermore, infrared scanning, radiometry and radar are used for detecting radiation temperature, subsurface temperature and mass structure differences.

The data rates during the acquisition are very high, ranging from 13 Mbit/s for 10 bands of multispectral scanning of a 10 x 1600 (with 8 bits) image up to 300 Mbit/s for a colour TV camera image of 600 x 800 (with 8 bits) image for each of the 3 colours. It is immediately clear from these figures that vast amounts of information are collected, sent to ground and stored, before they may be processed. The computer center on ground will have to perform a number of preliminary operations on the data before actually extracting the significant information. The main operations are: recording, converting, quality checking, formatting, correcting, storing (archives), reproducing (on paper), etc. (Refer to (4), Table II).

When we currently inspect a picture, perhaps the first thing we do is to notice that different objects are contained in it and that each one of these objects is made of parts which are grouped together in a region; sometimes these regions are sharply defined, other times, due to illumination variations, the edges are not distinct and some fuzzyness may be attributed to inter-region areas. The process by means of which we recognize different regions in an image is often called segmentation and, for most tasks, it is a rather fundamental one. In fact we want to be able to process only those components in an image which contain the significant information (for instance a road in an aerial photography, a telephone on a desk in an office scene, a chromosome in a microscope specimen etc.).

The interest in segmentation is not only due to the practical consequences of having a good algorithm to perform this task, but also because the different strategies which have been devised in the past years reflect the evolution of approaches to the conceptualization of image processing which contain alternative solutions to the problems of image representation, coding, storage, retrieval, labeling, etc. As we shall see, it is the way in which we view the problem which makes the difference in the solution found for that problem. The best way to deal with segmentation strategies is to start from the beginning: i.e. to give a reference framework from which the problem may be stated, approached and, why not, solved!

III. PURPOSE AND MEANING

Let us start with some definitions. According to (5) in many instances it is desired to obtain a description of a scene, a picture, in terms of its parts (regions or objects) and to generate this description it is necessary to segment the picture into these parts. This means that in order to analyze an image (regardless of the input source) it is convenient to eliminate from it the background and, to easily handle the remaining contents, to single out, in a sequential order, all the regions which must be measured in some useful sense. The extraction of these regions is generally named segmentation and the complexity of this operation strongly depends on the kind of regions to be segmented, i.e. on the problem at hand.

Many examples may be given of this task, the extraction of chromosomes from a histological specimen, the singling out of characters from a written text, the location of a cultivated area in an aerial photograph, etc. Another way to (6) define the purpose of segmentation on an image is to consider it as a subdivision of the image in a meaningful way to the viewer.

The fact of introducing "meaningful" as a necessary word to account for the explanation of this task, makes it impossible to have a general purpose analytical definition since constant reference is made to man and therefore to his purpose-oriented activity. In fact, many algorithms for segmentation are based on heuristics which are built on the evaluation of the meaning that some parts may have for the user. This meaning is difficult to establish once and for all and this explains why the technical literature is rich with approaches that try to imitate human segmentation by means of a computer in the least complicated way.

The difficulty of this problem is not only related to the subtleties which are embedded in any human procedure when deciding about the presence and location of the regions, but also and especially because the number of objective parameters which may be measured on an image is very high and none of them in general will be sufficient to ensure a correct segmentation.

To further stress the lability of any specific solution, we must also notice that the same observer may segment an image differently according to the purpose for which he is performing the segmentation, in a similar way we will look differently into a room if we are searching for a key holder and keys or if we only wish to check whether we left the light on in that room. Any search will be guided by the information we are looking for, and therefore some constraints will be forced on the strategy we are using for that search. A good illustration of this fact is the picture of a dalmatian dog, in which a dog will be seen only if one is looking for a dog, this fact corresponds to the use of a-priori knowledge for the search and, therefore, for the segmentation of the scene. (See fig. 1 from (6)). We now require some formal definitions in order to be accurate when describinbg some of the techniques which are used in segmentation.

IV. BASIC DEFINITIONS

Let us assume, as in (6) that the image is made of N by N elements which are obtained after a spatial and intensity sampling of the original picture, scene, etc. The result of this process (acquisition and digitalization) is a function f(i,j) which is non-negative and integer valued for every digital element (pixel) in the N by N rectangular array. The output from the process of segmentation is another array of the same number of elements (N x N) in which the elements will have a common label (value) if they belong to the same region (part, object) in the image. (See fig.2).

The regions will be defined in terms of a certain property which must either be constant or, at least, contained within a range of values. For simplicity, we will start by considering a uniformity predicate which is defined (6) as follows:

Uniformity predicate:

Let X denote the grid of sample points of a picture, i.e. the set of pairs.
(i,j) i = 1,2,3 ...., N, j = 1,2,3, ..., N
Let Y be a subset of X containing at least two points; a uniformity predicate p(Y) is one which assigns the value true or false to Y depending only on properties of the brightness matrix f(i,j) for the points of Y. (Moreover, if Z is a nonempty subset of Y then P(Y) = true implies always P(Z) = true.)

We now give some examples:
P (Y) = true, if the brightness value at any two points of Y is the same.

P (Y) = true if the brightness value of elements of Y does not exceed a certain maximum. On the other hand, P (Y) = true if Y has less than ten points is not a uniformity predicate since the true/false value of the predicate must only depend on the values of f(i,j) (brightness function). (See fig. 3)

We now give a formal definition of segmentation according to (6): A segmentation of a grid X, for a uniformity predicate P, is a partition of X into disjoint nonempty subsets $X_1, X_2, \ldots X_n$ such that

1) $\bigcup_{i=1}^{n} X_i = X$

$X_1 + X_2 + \ldots + X_n = X$

2) X_i is directly connected for $i = 1,2,\ldots,n$

$X_i \cap X_n \quad = \emptyset \quad \text{for } \forall i$

3) On each X_i the uniformity predicate $P(X_i)$ is true

$P(X_i) \quad = \text{true for } \forall i$

4) P is false on the union of any number of adjacent members of the partition

$P(X_i \cup X_j) \quad = \text{false for } \forall i \neq j \; (X_i, X_j \text{ are adjacent})$

This definition does not imply that only one segmentation exists nor that n is the smallest possible value for a partition.

We may also consider very simple cases in which the segmentation may be performed by simply thresholding the image on some brightness value B so that all those elements having $f(i,j) \geqslant B$ will become I-elements and all the remaining ones will become O-elements. As may easily be seen, this situation is seldom found in practice and only very trivial cases may be handled in this way. Since segmentation algorithms require the inspection of each point and of neighbouring points ("near" in some fashion to a central point) we must consider the "data structure" which constitutes the original image we wish to segment. For this reason we require the notion of a RAG (Region Adjacency Graph) which is homomorphic to the better known adjacency matrix. (Refer to figure 4).

Fig. 1 Dalmation dog

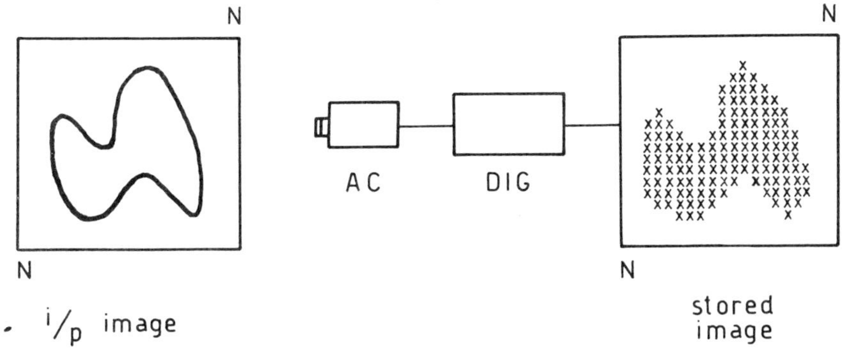

Fig. 2 Acquisition process

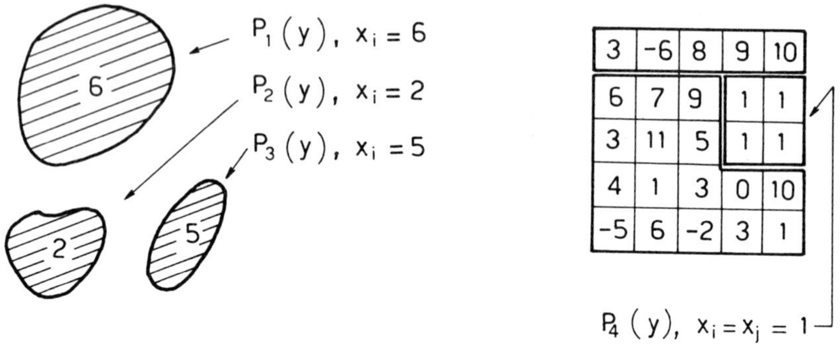

Fig. 3 Uniformity predicates

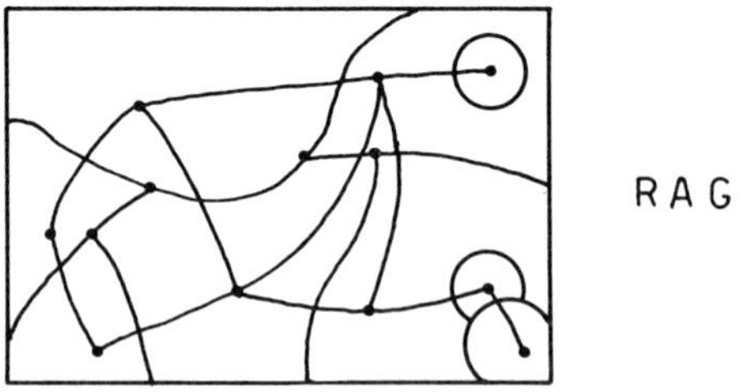

Fig. 4 Region adjacency graph (RAG)

In the shown image, a number of solid lines separates regions which have some common property, to each one of these regions a node is associated and a branch will join two nodes if the regions represented by the nodes are adjacent. The degree of a node corresponds to the number of branches that are associated to this node, for instance nodes of degree one correspond to holes (regions not connected to the external border of the image). The image frame will be represented by a node in much the same way as nodes represent loops in Circuit Theory and common circuit elements are shown by branches on the graph.

An image may be processed on a line by line basis (7) and each line may be segmented into a number of intervals so that a graph may be built by associating to each line interval a node and two nodes will be connected by a branch if the corresponding intervals lie on adjacent lines and their projections on a single line overlap. Refer to fig. 5. This particular graph may be called Line Adjacency Graph or LAG. It is important to note that adjacent intervals on the same line are not joined by a branch. In order to show the inclusion of regions instead of the adjacency another graph may be built, named the Picture Tree (PT). The root of this tree corresponds to the whole picture and the leaves to single pixels (refer to fig.6). If the grid has dimensions N x N, with $N = 2^L$, then the picture tree has L levels. Nodes at a distance j from the root correspond to squares of size 2^{L-j}; since every node in the tree corresponds to four pixels, the picture tree is called quartic picture tree (QPT). The importance of the QPT is due to its use in image segmentation where particular branches must be found in order to "cut" the graph according to some properties of the regions of the image.

A cut of a tree is a subset of its nodes such that no two nodes belong to the same path from the root to a leaf and no more nodes can be added to it without its loosing the previous property. Finally, a segmentation of an image always corresponds to a cut of the QPT as may be seen on the fig. 7 where the nodes belonging to the cut set are 1,2,5,6, b,9,10,13,14, d. The reason for introducing the graphs and the cut sets is not only formal but also practical since if an algorithm may be developed for extracting the cut set in a graph, this algorithm will also segment the image provided this image has been mapped into a picture tree. Many authors have devised representations of images which have a number of levels (in a cross section of a cone (8), in a pyramid (9)) and each level corresponds to a degree of coarseness (or definition) where some information may be better extracted. Normally these representations are called pyramid structures and it is interesting to notice that in terms of memory requirements, the increase in storage is of about thirty per cent of the one required for the whole image:

$$S_p = \sum_{j=0}^{L} 4^j = 1 \frac{1 - 4^{L+1}}{1 - 4} = \frac{4N^2 - 1}{3}$$

remembering that $N = 2^L$.

There are, as we have previously mentioned, a number of different approaches to segmentation and we may essentially distinguish between those that look for "common properties" (uniformity predicates) and those that look for "local differences" without excluding those that use both approaches to make the most of them. For the second approach in order to find "edges" or "boundaries", we now introduce an edge predicate as follows: if X denotes a grid of sample points of an image, an edge predicate Q(Y) is true on a non empty subset Y of X if: a) each point of Y has at least one direct neighbor outside Y and b) a uniformity predicate P is false over the union of Y and the direct neighbors of the points of Y.

Q(Y) = true if each point in Y has at least one direct neighbor outside Y and the difference between its brightness value and that of at least one of its neighbors exceeds a given value e. Furthermore, an edge, may be defined as a connected subset Y of X such that an edge predicate Q(Y) is true and there is no other subset Z of X which contains Y and for which Q(Z) is true, i.e. an edge is a maximal subset where Q(Y) is true. Unfortunately, once an edge is found, a segmentation is not always accomplished since edges may be discontinuous and therefore some areas will only be partially separated from the remaining components of the image. The edge extraction might be considered as a subproblem of segmentation and a great number of techniques have been developed specifically for this purpose, for a review see (10). In the local schemes each point will have a label depending on the value of its brightness and of that of its nearest neighbors and thresholding may be used to perform this. Pixels having the same value of brightness need not be connected and, in practice, different regions will appear which have a constant grey level. In the merging schemes the process starts from the pixel level (bottom-up in terms of the QPT) and if a certain property (uniformity predicate) is verified, then neighbors are associated to that pixel so performing a "region growing" operation (this name is generally used in the literature when referring to these schemes, see (11)). Edge growing is also based on the same principles although its target is edge detection more than segmentation.

An opposite scheme is the splitting one, which starts from the root of the tree (top-down in terms of the QPT) and subdivides regions where P(Y) is false, a merging is required since some pixels which are distant on the picture tree may be very close on the image. These schemes may use predicates over a large number of pixels (like the brightness variance) and this makes them less sensitive to noise. A combination of the above two schemes gives the split-and-merge technique (14) which reduces the computation time and does not increase the memory requirements. At the end of the application of these schemes the obtained segmentation is not suitable for human inspection and more processing is required using semantics and other knowledge about the specific problem, so producing human readable results.

V. THRESHOLDING AND EDGE DETECTION

As it was said before, thresholding is one typical example of segmentation and, perhaps, was the first way to produce an image where the objects to be processed were separated from the background. There are, indeed, many ways to threshold an image; this process may be described as follows:

$$s(i,j) = \begin{cases} 1 & \text{if} \quad f(i,j) > T \\ 0 & \text{if} \quad f(i,j) \leq T \end{cases} \qquad \text{for some chosen } T$$

and since each element is examined once, the time required is

$$t_1 = MN \text{ and the memory space required will be } s_1 = MN$$

The obvious problem lies inthe choice of the threshold T; for a bimodal grey level histogram (5) it may be considered to be in the valley of such a histogram. As it is known, a grey level histogram is obtained by plotting the number of pixels having a given grey level on the ordinate axis

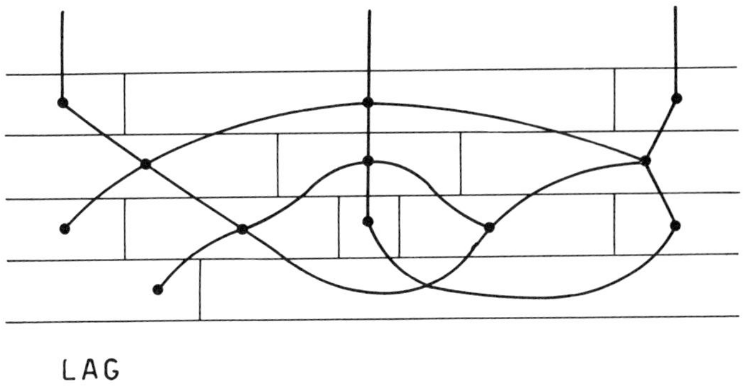

Fig. 5 Line adjacency graph (LAG)

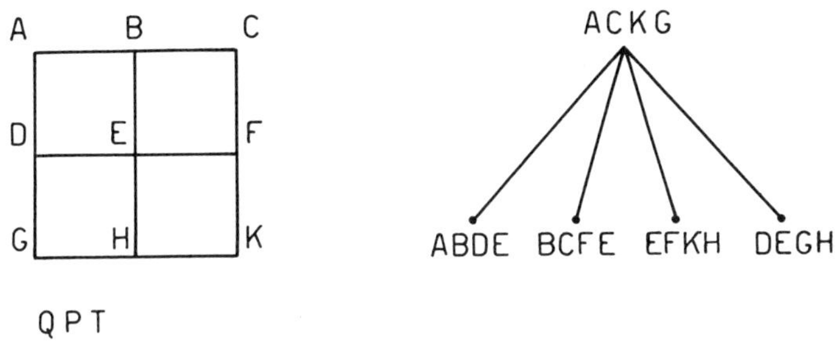

Fig. 6 Quartic picture tree (QPT)

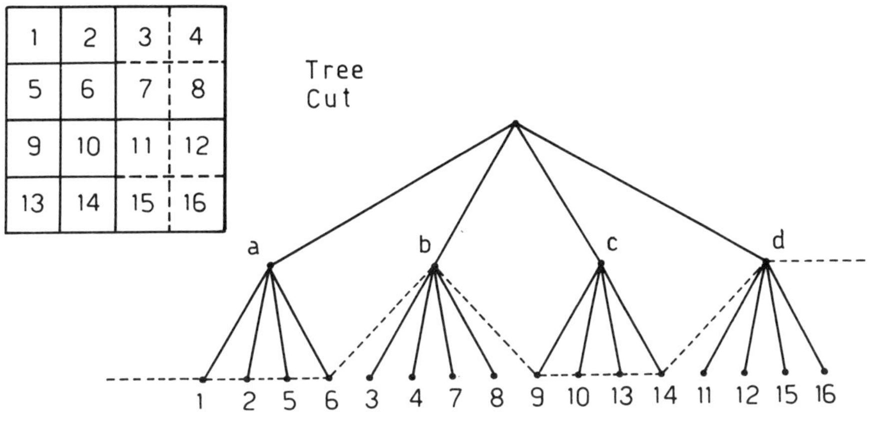

Fig. 7 Tree cut

and the grey level values on the abscissa axis, if the background is dark and the object to be segmented is light (rarely does this happen in practice in such a well defined manner) then it is very likely that two peaks will appear in the histogram, one for the background and the other for the object. Refer to figure 8 where a bimodal grey level histogram is shown. One possible way to force the histogram to become bimodal is by means of an algorithm (12) which is given below:

1) for each (i,j) compute the gradient norm by taking the average g(i,j) of the absolute values of the differences between f (i,j) and f(i+1,j), f(i-1,j), f(i,j+1), f(i,j-1) respectively (the chosen neighbors are those on the same column and row). If g(i,j) exceeds a threshold H, set h(i,j) = 1 else h(i,j) = 0.

2) form the histogram of h(i,j) xf(i,j) and choose a threshold T at a valley of the histogram. (If no such valley exists the method fails).

3) perform a threshold operation on f(i,j) with T to obtain s(i,j).

This method works in general since it is easier to find the threshold H than the threshold T; unfortunately the computational requirements and the memory space are higher than in the previous method. For the gradient (one subtraction, one absolute value and one addition for each of the four neighbors for all the elements in the image), then a comparison with threshold H, then the formation of the histogram ($\log_2 L$ comparisons for L grey levels and one addition for all the elements in the image) and the valley detection with the final thresholding, are all the elementary computations required for this method. Furthermore two matrices are necessary, one for h(i,j) and the other for s(i,j). We then have $t_2 = 4.3.MN + MN + (\log_2 L) MN + 2L = (14+\log_2 L) + 2L$ and $s_2 = 2MN$.
For a comparison between a seguential and a parallel thresholding procedure see (13). Thresholding may also be performed on different thresholds along the grey scale or on different parameters like the three primary colours, the average of the three colours, hue, etc. Small regions of the image may be analyzed for detecting the valley in the local histogram and a number of variations on this theme may be thought for improving the efficiency of the thresholding technique for segmentation, in some cases semantic information may also be used to improve the results and to avoid loosing contours of regions with constant brightness variation or considering noise as defining local regions to be segmented. Edge detectors are also designed for segmentation purposes although the techniques are different since a local variation of brightness is considered as the main cue for the location of the fragment of an edge. For instance, the following local operation may be considered a basic one in this field.

$$g(i,j) = |f(i,j)-f(i-1,j)| + |f(i,j)-f(i,j-1)|$$

A very popular edge detection algorithm is due to Sobel and may be written as follows:

$$g(e) = \left| [f(a)+2f(b)+f(c)] - [f(g)+2f(h)+f(i)] \right| + \left| [f(a)+2f(d)+f(g)] - [f(c)+2f(f)+f(i)] \right|$$

where the meaning of the letters is given in figure 9. If the image is binary and a corner is centered at e, we will then have g(e) = 6 whilst if a horizontal edge is present in the image we will then have g(e) = 4.

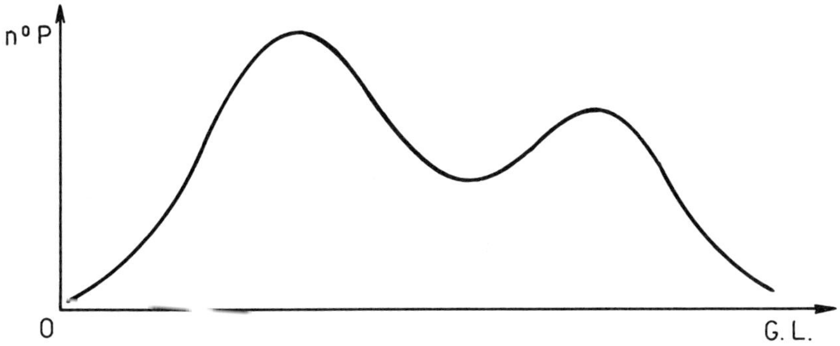

Fig. 8 Bimodal grey level histogram

a	b	c
d	e	f
g	h	i

Fig. 9 Sobel's nomenclature for neighbors

In general local gradients on the discrete image may be described by the formula

$gh(i,j) = \sum_k \sum_\ell \; h(k,\ell)\; f\,(i+k, j, +\ell)$ for $h\,(k,\ell)$ a weighing function. A large number of algorithms have been suggested for accomplishing edge detection satisfactorily and each author has his own preferences which largely depend on his own experience and on the kind of images he has been processing.

VI. SPLIT-AND-MERGE SEGMENTATION

From a more philosophical angle, the use of semantics for achieving good segmentation has been both suggested and used by a number of authors. It is still an open question whether it is better to introduce semantics at an early stage of the segmentation task and therefore get involved with a larger computational burden or if it is perhaps better to use it at a later stage for substantially improving the results obtained only on the basis of the segmentation algorithm. Lastly, it may sometimes be unnecessary to use semantics at all if the problem is not too involved and, or, if the procedure for segmentation is powerful enough.

As mentioned before, splitting-and-merging tries to make up for the disadvantages of each technique on its own and essentially relabels an initial set of regions on an image until some (formal segmentation) properties are satisfied.

Since an image may be decomposed in "quarters" (see the Quartic Picture Tree introduced above) every single pixel may be grouped together with other three adjacent pixels (the four of them will be the leaves of a QPT) so as to become one region, this process is referred to as merging. On the other hand if coming from the root of the tree we subdivide a region into four elements, this is referred to as splitting. In both cases we will merge or split provided a certain predicate is true and when all the merging and splitting is accomplished (the convergence of the method may be proved) there is still need for a final procedure to eliminate small regions which are generally created in the process. The new predicate will not only contain values of f(i,j), but also values oof the areas in which f(i,j) is computed, in this way, a smaller number of regions is obtained and the final segmented image is easier for human interpretation.

Returning now to the formal definition of segmentation, which is given in section 4, we should note (14) that merging starts with a partition satisfying 3) and proceeds to fulfill 4) whilst splitting starts from a partition satisfying 4) and proceeds to fulfill 3). A split-and-merge procedure starts with an arbitrary partition which satisfies neither 3) nor 4) and ends satisfying both. In order to achieve segmentation, the PAT (picture adjacency tree of the image) is considered so as to extract the cutnode (see definition above) which is the only part of the graph which requires to be stored in memory and all the new arrangements of the nodes will modify the cutset until the last obtained cutset provides the final segmentation.

Let us refer to fig.10 (from (14)) where an artificial image is shown together with a hypothetical segmentation into regions labeled with numbers of one, two or three digits according to the level in the tree to which they belong in their node representation. The root of the tree represents the "external world" or the frame (sometimes labeled node 0), node 4 has no descendents, nodes 1, 2 and 3 have a number of descendents shown on the figure. In this tree the leaves are blocks

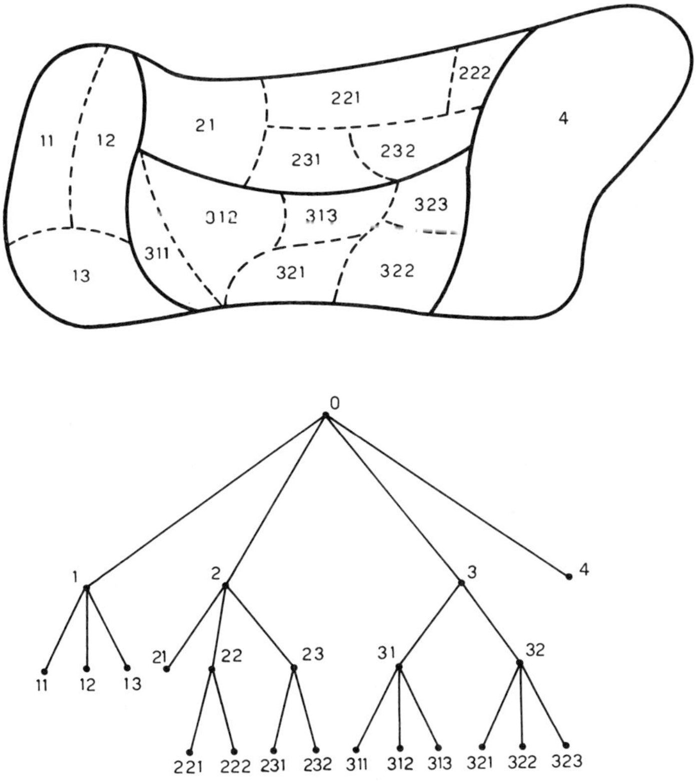

Fig. 10 Artificial image: each number corresponds to a segmented region.

of one element (pixels) and the other nodes at level L, named b, represent a block of elements having sides of length $n/2^L$. The element is positioned on the top left corner of such a square and will have coordinates i,j. The brightness function f(i,j) will be associated to each node with its maximum and minimum value for the corresponding block.

Let us now see how is the merging performed between adjacent blocks. As mentioned earlier, two regions should be merged if they share some common property, for instance if the difference between the maximum of f(i,j) and the minimum of f(i,j) is below a certain threshold, then the components of the block (which might be other blocks) should be merged, this will be done on all the levels of the tree, level by level for all blocks. On the other hand splitting will be performed inside a block if the difference between the maximum and minimum value of brightness exceeds the given threshold. Merging may proceed until only the root itself (and therefore it will terminate) and splitting may only proceed until all the leafs are obtained and, again, will terminate; furthermore the two processes are mutually exclusive since once merging has been done and followed by splitting no other legal partition might be achieved (no merged blocks may be splitted and no splitted blocks may be merged again).

VII. SEGMENTATION OF A SUBURBAN AREA

Whilst the formal approach to segmentation we have seen so far is extremely general and may therefore constitute a framework in which the continuity predicates may be changed until satisfactory results are obtained, in real life images, the complexity is still too large to be harnessed within graph theory and a number of active workers in the field have tackled the problem from a different angle. For instance, in the field of aerial photographs, the size of the data is really very large, the conditions under which the photographs are taken differ widely according to time and season and a model of what will be seen by the camera is very difficult to imagine. It is because of these reasons that although many measurements on these images were made in the past (spectral intensity on different wavelengths) the results of classifying the points by maximum likelyhood (supposing normal distribution for the samples of each category) were not accurate enough. Mostly because of the influence of the AI community on the PR community, other measurements and features were introduced like the local structures in the image and the relationship between these structures, namely the syntactic approach. Lines are extracted so that all the line-like objects could be tested for rivers, roads, piers etc. and at the same time rectangular-like objects could be checked for houses, city blocks, cars, ships, etc. Again the problem of the great amount of structures to be tested against a very large data base coming from the image implies expensive computations and therefore long execution times. At this point we are near to round-circling reasoning: if there are many bits to analyze we must necessarily take a long time, if we should only analyze the significant ones we should also find out which are the significant ones and therefore analyze them all...... returning to the beginning.

Perhaps we might reduce the regions we want to analyze by firstly detecting some gross features of those regions (identifying them without accurately classifying the contents of the region) and this should be done by using all the available information on the target itself. At the same time we want to use information about the environment of the target, like roads and parking lots if the target is cars, or water if the target is ships and so on.

This approach is based on common day experience about how we, humans, look at a photograph when we are motivated at finding a specific aobject in the image. This process may be called "focus of attention" (15) and does not require, at least at the beginning any preliminary knowledge about the objects that we are looking for since some regions which appear very different - clear and distinct from the background for instance - will immediately call our attention. Once we have focused on a part which seems interesting to us we will then use all the available information on the objects we are looking for so as to guide the search and use short feedback loops to correct small errors and improve the reliability of each stage within the process. Along this line, we could include merging algorithms which start by considering a smaller number of elements in the image (for instance by local averaging and shrinking from the edges inwards in a pyramidal way) as mentioned earlier. Some of the more evident features may be easily extracted on the smaller image and in (16) a method called "edge preserving smoothing" ensures the smoothing of the flat areas and the sharpening of the blurred areas. This method is based on the computation of the variance of the intensity from an element with respect to its neighbors within a rectangular window; in this way the rectangle having the smallest variance of the intensity change is found and the average intensity value is given to the central element in the newly computed image. This is done on all the elements of the image a number of times until no more significant changes occur thus obtaining a better quality image. Let us now review the basic steps of the procedure (17) which was adopted for the analysis of aerial images:

1) Firstly, strong and typical features are extracted from the aerial photograph (focus attention)

2) Further analysis is performed on the extracted portions by using some knowledge of the objects

3) Some short feedback loops have been originated at step 2);

typical objects which are prominent are first extracted by the default parameter values. Moreover the parameter values are actually computed from the extracted objects and the initial default parameter values are replaced by the extracted parameter values, object detection is repeated for new other objects present on the image until no more objects are found. The focus of attention part of the process is based on the detection of the following features:

a) large areas of uniform smoothness (vegetation area, water surface)

b) long thin regions (roads, rivers, railroads)

c) stable color features (green for forests, blue for water)

d) heavily textured areas (forest, vegetation area, urban area)

Many methods may be used for the extraction of the features mentioned above and, specifically, for a) simple thresholding using a double threshold for establishing a level of smoothness, for b) Fourier analysis and low pass filtering to eliminate frequent textures and for sharp energy concentration we find rivers and stick-like objects in general. Using the spectral intensities of RGB and IR we may start segmentation by thresholding. Saturation and hue information can also be used for improving region segmentation.

For some specific texture parameters the technique known as shrink-expand may be used (5) where gaps are filled and notches are cut off by means of an isotropic contraction and expansion of the objects present on the image. In order to better follow the overall process applied on the aerial photographs we will now refer to figure 11 in (17). (The flow chart has been redrawn for clarity in the context of this lecture).

The analysis may be divided into two parts: task independent and task dependent; the first one is made of three steps; a) smoothing - b) segmenting - c) further segmenting; a) uses the edge preserving smoothing mentioned above for improving the image quality; b) uses multi-spectral information for extracting homogeneous regions (the continuity predicate uses spectral intensity as the argument) and c) the segmented regions are further analyzed so as to extract cue regions.

The next part of the process is task dependent and each type of cue region is analized so as to extract parameters of the object by the specialized object subsystem and then the properties of the found objects are stored in a common area of the system building up the property table and feeding back to the corresponding object subsystem. Five kinds of regions are extracted:

1) large homogeneous regions
2) elongated regions
3) shadow and shadow-making regions
4) vegetation regions
5) high contrast regions

The method has been applied to analyze the landscape of a suburban area in Japan. The photographs have been digitized on a 256 by 256 array of elements with 256 grey levels in each of the three wavelengths of the spectrum (RED, GREEN and BLUE) plus the IR (INFRARED) wavelength. One pixel of the digital picture corresponds to 50 cm by 50 cm on the ground, considering the different spectral readings, the input to the computer is made of four digital images.

After preprocessing the image with the smoothing algorithm briefly sketched above, each image in the 4 wavelengths is partitioned with a region growing or merging algorithm which may be described as follows:

1) if all pixels are labeled, then end; else take an unlabeled pixel and assign a new unused region number
2) if the differences of grey levels in the 4 wavelengths between a new labeled pixel and its neighboring pixel (4 adjacency is assumed) are less than thresholds (for i=R,G,B,IR respectively), then merge the neighboring pixel and asign the same region number
3) iterate 2) until no two adjacent pixels are merged
4) go to 1).

This merging algorithm seems too simple to work and therefore one would expect mismerging, but some preliminary measures have been taken to avoid such mismerging. In fact, the initial smoothing avoids the effect of blurred edges in the computation of the merging algorithm, the threshold is adaptively chosen for each wavelength on the data and gentle slopes in grey levels (on all wavelengths) are neglected (typically due to shading effects). The threshold computation is performed by means of an algorithm similar to the one which has been previously illustrated since the valley is sought not on the grey level histogram but on the differential value histogram.

The image is grouped into 16 blocks since if all the image would be computed at once, the valley would not show very clearly because differently textured regions would average out and produce large populations for many differential values. Since smoothing is applied independently on each wavelength, some boundaries differ slightly on each of the different images taken on the same landscape and therefore small regions, distant less than four points from a large region having the most similar multispectrum property, are merged so obtaining an image called Label Picture and on it the Property Table is computed on the basis of average grey levels value, size, location, shape of the detected region together with a region number.

Three basic shape features are now computed, the first (FIT) is the ratio (in terms of area) between the region area and the area of a rectangle (minimum bounding rectangle) when the main axis (parallel to one couple of sides of the rectangle) is rotated by $\pi/4$, $\pi/2$, $\pi 3/4$. The second feature is the ELONG or elongatedness of the region as the ratio between the long and short sides of the minimum bounding rectangle and the third is the direction of the long side of the minimum bounding rectangle, DIREC is an angle.

We may now remember that the problem of measuring the shape of a digital object is still an open one. One very well known approach is to consider the convex hull of an object after defining the convexity property on the discrete. From this idea, conceived in a sequential manner, the minimum perimeter polygon was suggested (MPP) (18); essentially it might be imagined as the shape obtained by putting a rubber band around the contour of an object, convexities will produce vertices of the MPP and concavities will remain inside the MPP. Unfortunately, the computational requirements of such approximations are not small enough to become practical and, furthermore to take full advantage of the MPP the list of vertices must be retained so that no "shape factor" might be produced to globally characterize the shape. Many other approaches to find significant points (critical points) along the boundary of an object have been suggested in the literature. On the global approach to the quantitative evaluation of shape the best known parameter is compactness, which is given by the ratio of the perimeter squared over its area P^2/A. For a recent comparative review on this subject see (19).

As mentioned earlier, the next step is dedicated to the extraction of cues from the detected regions and using the properties of size, shape brightness, color and texture, five cue regions are sought: 1) large homogeneous regions), 2) elongated regions, 3) shadow and shadow making regions, 4) large vegetation regions and 5) high contrast regions. We now give some hints for understanding the strategies employed in cue region extraction. The main point is how to decide on the value of "large" regions independently from the altitude from which the photograph was taken.

Firstly, the histogram of areas of homogeneous regions in the LABEL picture is built and the valley of such a histogram is found by the method mentioned above for valley detection. Once the threshold on the area is established, all areas above such a threshold are extracted. The elongated regions are detected after considering those regions with ELONG greater than 3.0 but since curved regions may also be elongated, a special procedure is developed for the detection of such curved regions:

extract regions with FIT less than 0.5 and if they have holes, fill them in. Then extract the longest path on the thinned region (using a special algorithm for longest path detection). If r_i ($i = 1,2,\ldots,N$) denotes the successive position of points on the longest path, compute the

O.P.	ORIGINAL PICTURE
SM	SMOOTHING
S.P.	SMOOTHED PICTURE
SEG	SEGMENTATION
L.P.	LABEL PICTURE
B.P.T.	BASIC PROPERTY TABLE
EX.C.REG.	EXTRACTION OF CUE REGIONS
L.H.R.	LARGE OMOGENEOUS REGIONS
E.R.	ELONGATED REGIONS
L.V.R.	LARGE VEGETATION REGIONS
H.C.R.	HIGH CONTRAST REGIONS
P.T.	PROPERTY TABLE
EX.OV.REG. ...	EXTRACTION OF OVERLAPPING REGIONS
M.B...........	MESSAGE BOX
H.C.V.R.	HIGH CONTRAST VEGETATION REGIONS
C.F.D.R.......	CROP FIELD DETECTION ROUTINE
R.D.R.	ROAD DETECTION ROUTINE
G.L.D.R.	GRASS LAND DETECTION ROUTINE
W.D.R.	WOOD DETECTION ROUTINE
H.D.R.	HOUSE DETECTION ROUTINE
O.T.	OBJECT TEMPLATE
F.P.R.O.	FEEDBACK OF PROPERTIES OF RECOGNIZED OBJECTS
O.P.T.	OBJECT PROPERTY TABLE

Fig.11 Block diagram of the segmentation process

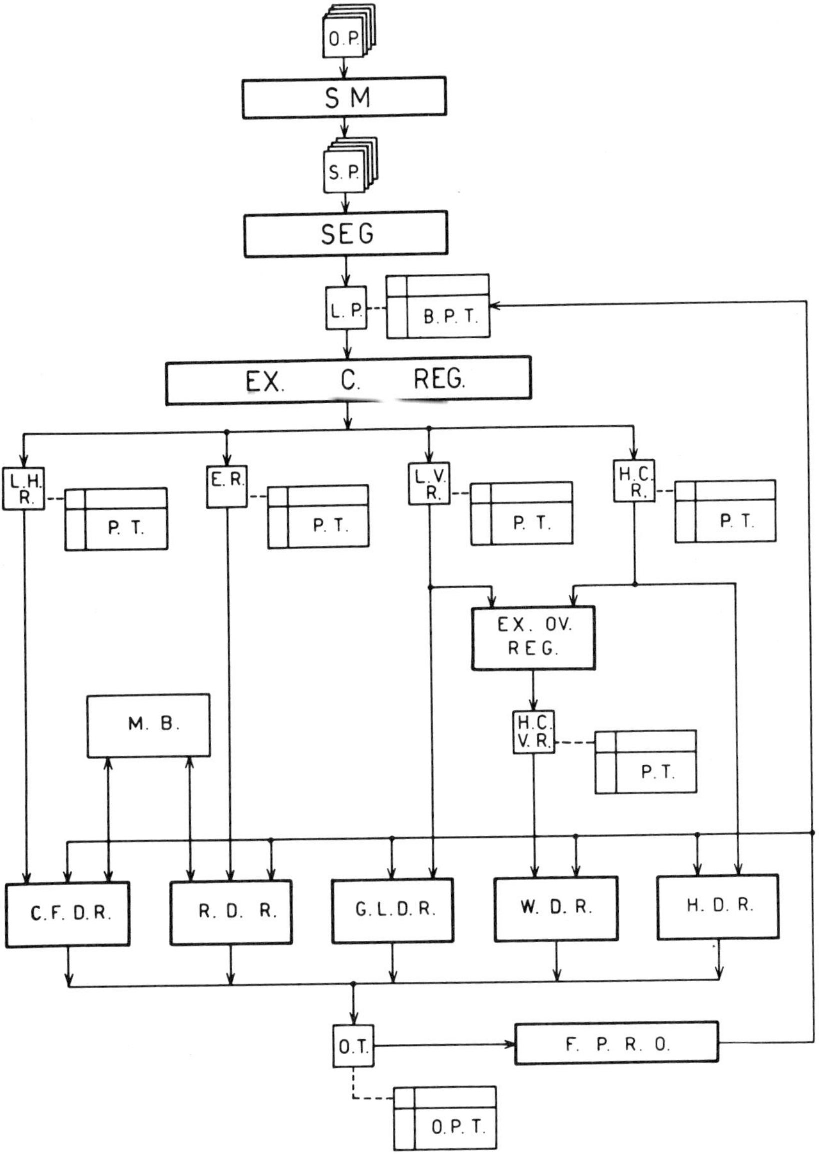
O.P.
S M
S.P.
SEG
L.P.
B.P.T.
EX. C. REG.
L.H. R.
P.T.
E.R.
P.T.
L.V. R.
P.T.
H.C. R.
P.T.
EX. OV. REG.
H.C. V.R.
P.T.
M. B.
C.F.D.R.
R. D. R.
G.L.D.R.
W. D. R.
H. D. R.
O.T.
F. P. R. O.
O.P.T.

Figure 11

width w_i by the following formula $w_i = n_1 + \sqrt{2}\, n_2$ where n_1 denotes horizontal or vertical steps and n_2 stands for diagonal steps along a direction normal to $r_{i-1} - r_{i+1}$. Compute the average width of the region,

$$W = \frac{1}{N-2} \sum_{i=2}^{N-1} w_i$$

finally extract those regions whose $ELONG2 = \frac{\text{length of longest path}}{W} > 2$

The extraction of shadow and shadow-making regions is based on the difference of brightness values, obtained for shadowed areas, between the R, IR wavelengths and the G, B wavelengths. In order to enhance this difference, which is due to the scattering of short wavelength light in the atmosphere, the value of brightness I will be weighed as follows: $I_W(i,j) = 1/6(2I_R(i,j) + I_G(i,j) + I_B(i,j) + 2I_{IR}(i,j))$. Next, a histogram of the grey levels of the new variable brightness is built ensuring a value of I_1 which neatly separates the histogram into a minimum brightness part and a part above the average brightness level.

We now proceed for the shadow-making regions which are adjacent to the shadow regions in the direction of the sun (known from the flight information) and have a long common boundary with a shadow region in the counter direction of the sun. These regions, shadow and shadow-making, are input to the property table of LABEL picture and will be later used for object recognition.

The large vegetation regions use information from the different spectrum bands since the value of reflectivity of vegetation is known to depend both on wavelength and on the time of the year. The chosen parameter will be $I_{IR}(i)/I_R(i)$ where $I_{IR}(i)$ and $I_R(i)$ are the average brightness values over a region i. Since house roofs have the same spectral response in the R and IR wavelengths, there is the need to consider the B band since roofs may be discriminated from vegetation by thresholding on a level of the blue band, only low brightness in blue excludes roofs (since roofs have a high value of blue intensity). Once the first vegetation areas are found large vegetation areas may be obtained by merging provided the ratio $I_R(i)/I_{IR}(i)$ is above 1.4.

The last regions to be extracted are the high contrast ones where trees, houses, roads and shadows may be present but which may not be recognized only on their features. This class of regions is made of lots of small areas which need to be further investigated by the next part of the process. The method for extracting these regions is the following: extract the boundaries of the homogeneous regions in the LABEL picture which has more than half its area contained in the regions extracted at the previous step.

The large homogeneous regions previously extracted are excluded from this process. These high contrast regions are registered in the high contrast region mask, the intersection between the high contrast region and the large vegetation region is registered in the high contrast vegetation region mask.

We now proceed to the third and final part of the process which deals with the analysis of the cue regions. As mentioned earlier, there is a subsystem for the recognition of each object supposed to be present in the image. The input data to the subsystem is made by the basic property table, the LABEL picture, the object property table, the object template, the cue region mask and each region property table. When analyzing large homogeneous regions (which may be a crop field, a bare soil region or a grassy area) the boundary is considered (since a crop field has a given boundary) and analyzed to check for straightness of boundary edges. This is performed by the following steps:

For each point Pi (i = 1,2,...,N) on the boundary, the angle θ_i is computed between two straight lines connecting P_i with P_{i-n} and P_{i+n}, if $\hat{N}$ denotes the number of points for which $|\theta_i| \leqslant 22.5$ and if $\hat{N}/N > .6$ then this region will be considered as having a straight boundary. Some regions may be both recognized as being elongated and largely homogeneous so that the final decision is suspended until the subsystem analyzing elongated regions is not able to connect them to other elongated regions. Crop fields are classified in two groups according to the presence of vegetation which is checked by means of the property table.

The elongated regions include rivers, roads and since the covered area (by the photograph) is rather small, we may assume that these regions intersect the frame of the image at some angle. Sometimes these regions are broken down by the presence of cars on roads or bridges on rivers or shadows, therefore some means must be devised for merging the different components of the elongated regions. The properties of elongated regions are: no vegetation, not shadow region and not shadow-making region so that these properties are firstly checked on all the elongated regions found until now. Then the end point of the longest path is found together with the width of the candidate region, for each pair of candidate regions the following values are checked: the differences of the average grey levels in the four spectrum wavelengths are tested to see if they are smaller than θ_i (i=R,G,B,IR) respectively; the ratio of the widths of two regions to see if it is between 2/3 and 3/2; the shortest distance between end points of two longest paths to see if it is smaller than 3 W_{min} (where W_{min} is the minimum width of the two regions) and finally the difference of the directions of the two regions to test if it is smaller than 45° . If all these conditions hold, then the two candidate elongate regions are merged. Next, those regions whose ends touch the frame of the image are considered as roads and the regions which satisfy the following requirements are merged:

their hue differences are small, the ratio of the widths of a region and roads is between .5 and 2 and the distance between one end point of a region and the recognized road region is less than 3 W_{min}; this last step is repeated if a new region is connected to another elongated region. In this way a network of roads may be recognized.

A high contrast vegetation region will be confirmed to be woodland if small element regions in a high contrast vegetation region have irregular shapes and if these regions contain shadow and shadow-making regions. The analysis of a residential area includes the detection of shadow and shadow making regions within the high contrast regions and the detection of small constitutent regions with rectangular shape. The difference in direction of boundaries os small objects is near to 90 (houses). Now, if the above properties hold, we have a residential area in which rectangular shapes are detected (FIT greater than .65); shadow making region, no vegetation. So far, in the examined image no region satisfies the properties for residential areas.

It is interesting to remark that when some wood and road regions are detected and removed from the high contrast region, the properties of residential areas become prominent; on the remaining area of the high contrast region the "houses" program is run and some houses are detected, the properties of these houses (spectral and morphological properties) are stored in the object property table. These properties are used to modify the "houses" program so that now we incorporate other cues to enable a complete extraction of houses. These other cues are: similar spectral response to other previously found houses, adjacent to other houses within a shadow making region, rectangular shape (lower FIT value).

As was seen in this description of Nagao's work the basic idea of a flexible segmentation scheme is to use all the knowledge which is available before the image is analyzed and then during this analysis new knowledge is incorporated as soon as it may turn useful. This program is written in Fortran, runs on a FACOM M-190 computer and the total running time including the input-output of image data is of about 10 minutes. The importance of the pre-processing algorithm (edge-preserving-smoothing) should not be underestimated since the quality of the data to be processed is very important for finding the correct boundaries of the regions in the image.

VIII. PARALLEL NUMERICAL COMPUTATION

Many years ago, parallel processing, was used in connection with the possibility of executing arithmetic operations on a full word rather than one bit at a time and, more recently, the presence of many processors within a computation system enables the simultaneous execution of different instructions. Two different architectures may be considered having a certain amount of processors: one in which all the processors execute the same instruction (on different data) and another, more complex, in which all processors execute different instructions again on different data, the first kind is known as SIMD (20) class of machines (for Single Instruction on Multiple Data) whilst the second kind is called MIMD class of machines (Multiple Instruction on Multiple data).

The purpose of using SIMD and MIMD machines is to reduce the computation time so that a certain speed up factor may be defined, possibly in a way which is independent of the task. According to (21) one possible means to define the speed gain obtained by using a machine with p processors is the following: T_1 is the number of unit time steps required to perform a computation with a single processor machine and T_p is the number of unit time steps with a p-processors machine so that S_p (or speed up gain) will be $S_p = T_1/T_p \geqslant 1$ and an efficiency (using the p-processors) is $E_p = S_p/p \leqslant 1$. This ratio might be considered as the real speed gain divided by the theoretically maximum speed up (p).

In arithmetic we may parallelize the computation of expressions by simply grouping the "atoms" of the expression in pairs and reducing the height of the tree which represents the different computation steps. If we start from an 8-atom expression (the atoms may be connected by any one of the four arithmetic operators) we will obtain, at each step, a reduction to half the number of atoms, i.e. 8,4,2,1 so that only three computation steps are required (instead of 8). In general, for a base 2 logarithm, the number of steps required for the computation of an arithmetic expression having n atoms will be proportional to log n. This is very nice when n is large but unfortunately the number of atoms in these expressions is not very high, it is generally of the order of five or six. On the other hand the images that require processing can be thought of as an ordered set of atoms, (better known under the name of pixels) on which some computations must be performed so that their number is large (from 10^4 to 10^6) and known a priori.

As it is discussed in detail in special survey papers (22) not only must we try to use all the available processors all the time in order to achieve maximum speed up but we should also try to increase the through put between memory and each processing element and, in more sophisticated architectures, allow for any processor to access any memory element. From all this it is immediately clear that the advantages to be gained simply by increasing the number of processors are obtained up to a certain extent in a limited number of cases; in other words, the speed gain does not increase indefinitely as we increase the number of processors.

IX. SOME PARALLEL MACHINES (23)

Since the first paper model of a parallel computer (spatial computer (24)) for image processing was suggested in 1958 a number of projects were started and machines were built for implementing the main ideas contained in the "spatial computers". The oldest, and perhaps the best known of these projects, was the one named ILLIAC in the fifties which developed into the ILLIAC IV. This machine is basically made of 256 processing elements subdivided into four 8x8 arrays called quadrants. Actually, only one of these quadrants has been completed, to our knowledge. The PEs are four connected (lateral and vertical neighbors only) and may access their own 2K words of 64 bits each. A control unit is responsible for driving the PEs according to the instruction which is decoded, and all PEs either execute, or remain idle, for that instruction. The memory cycle is of 250 msec. and the whole machine is hosted by a Burroughs B6500, it may be programmed in its own rich Assembly language but a number of other languages - nearer to the programmer's level - have been proposed like TRANQUIL, IVTRAN and, more recently, GLYPNIR. The typical applications of ILLIAC IV include matrix algebra, linear and integer programming, weather forecast and modelling digital filter design.

From a different angle and mainly based on the concept of associative memories, the most important machine is STARAN, designed and built at Goodyear Aerospace. This machine was designed to handle vast quantities of data at high speed and therefore has efficient retrieval facilities which are based on the "contents-addressable" mode typical of an associative memory structure. The basic processing element contains a memory of 256 words of 256 bits each (9216 words in the most recent STARAN) and a permutation network which allows different data organizations and displacements within the associative memory; the maximum number of PEs is 32 and depends on the specific installation. The STARAN is hosted by a CDC 6400 and includes a number of special peripherals. The basic clock cycle is of 100 nsec and the main applications are in map processing (particularly in the area of radar maps). This machine uses an Assembly language called APPLE which is procedural and extensible according to the specific user needs.

Still from a different angle, a new machine called PEPE, has the main feature of allaowing the attachment of a variable number of PEs. Each PE is made of an ALU and a memory of 1024 words of 32 bits, a correlation unit and an associative output unit. A main controller, called PICU (Parallel Instruction Control Unit) is microprogrammed to drive all the present PEs. The clock cycle of PEPE is 100 nsec and the machine is hosted by a CDC 7600 and its main applications are of military nature. This machine may be programmed in PAL which is its own parallel assembly language.

A machine which has been developed in 1972 in ELSAG (Genoa, Italy) called EMMA (Elab. Multi Mini Associativo) is based on a hierarchical system of families each one having a bus with a cluster of PEs up to a maximum of 128 units with a memory capacity of 2 M words of 16 bits each. The system grows according to the application and each PE communicates with the others and with the main controller which is a standard 16 bit word minicomputer. The machine cycle is of 1 μsec. for transferring data and of 0.6 μ sec. for executing a basic instruction. The application of EMMA include OCR for the Post Office and for meter readings. The language used is a macroassembler named AUO.

So far we have briefly enumerated a number of machines, typically SIMD machines, which are only loosely related to image processing but which constitute a rich background of experience both in the testing of different computer architectures and in the experimental evaluation of the relative computational efficiencies on a varied range of applications.

In connection with image processing a number of machines were also developed which either incorporate a true full parallelism (as in the CLIP (25) family of machines) or a partial parallelism which is then used for scanning the array to be processed in order to combine high speed with some of the more convenient computer hardware circuitry; (as in the PICAP (26) machine or in the commercially available diff.-3 (14)). In the first case we have a true SIMD machine which has, in its latest version CLIP 4, about 10^4 processing elements with eight connectivity (or, as an option, six connectivity), a memory of 32 bits for each PE and an external control which decodes the instruction and drives the PEs accordingly. A clock cycle of 2 μsec is typical of the MOs technology used in CLIP 4 but other functions like propagation of signals from a PE to neighboring PEs take much less. The CLIP 4 assembly language is called CAP 4 and easily allows to write and read a program in terms of local neighborhood operations (see below). On the other hand, PICAP is parallel within a 32 x 32 array which scans the larger (256 x 256) image at high speed. This machine may be programmed on a high level language, FORTRAN, by means of the use (by CALLS) of specially written subroutines which perform the local operations on all the elements (pixels) on the array.

X. SEQUENTIAL VS. PARALLEL IMAGE PROCESSING

One possible way to compare the relative merits of the use of parallel machines for computing images is to select a set of typical tasks performed on images, for instance see (27) and then to code programs in the corresponding assembly languages both of a serial computer (we have used a HP 2116B) and of a parallel computer (we have considered CLIP 3). Firstly the best algorithm to perform the task is chosen, so as to match the computational procedure with the hardware that will implement it, we then count the number of clock cycles which are required on the two machines for the same task and plot their ratio in terms of the linear side of the array. This side length also corresponds to the square root of the total number of processors which are contained in the parallel machine. This comparison was made for the following tasks: smoothing, thinning, binary contour extraction, perimeter evaluation and thresholding. As in all evaluations of this kind the choice of the tasks is crucial. The first three, which apply to binary images, are particularly suited for parallel processing, since they require the computation of a value on the basis of the values of all the elements contained in a 3 x 3 neighborhood of a central 1-element. The interesting fact is that when perimeter evaluation is performed, which involves a numerical computation, or a thresholding which is much more arithmetically based, we still have considerable gains as may be seen from the graph. In particular, thresholding (28) requires the construction of the grey level histogram, the detection of a valley on such a histogram and finally the comparison of the value of each pixel with the one established as threshold which is obviously done in one single step: all elements at once becoming I-elements if equal or greater than threshold. From the graph, we may see that the speed gains vary from 10^1 to 10^6 for a typical number of 50 pixels for side image, and for different tasks. From this analysis it may be concluded that it is difficult, from a general point of view, to establish the speed up gain in a multiprocessor machine when handling numerical computations, and lead to believe that "the cost effectiveness can be increased if the parallel processors are used at the algorithm level more than if parallel gates are used at the arithmetic level" (21).

XI. CELLULAR LOGIC OPERATIONS

The reasons for wanting to perform parallel processing on images have been already stated above and are essentially based on the increased computational speed that may be achieved in this way. Within the vast class of operations that can be performed on an image, a basic type of operation is represented by one called Cellular Logic Operation (29) which is frequent in practice and may be directly implemented on a CLIP machine or on any other array processor of the SIMD class.

Basically the CLO is represented by the application of the next state function F at t=0 on the internal states of all PEs, I^t_{xy}, where the superindex indicates a discrete time value and the subindexes indicate the coordinates of each PE. The internal state of a PE depends, in this formalization, on five independent variables, namely $I^t_{xy} = ((a,b,c)_{xy}, w, z)$ The first three variables might differ from element to element whilst the other two are equal for all the PEs and may be considered as instructions. The variable "a" represents the binary value of the pixel in xy of the input image, "b" is the value of a corresponding pixel xy of a second image (which is not necessarily present in every computation), "c" is a function of the output value of the neighboring PEs (or of a subset of these neighbors) and a given threshold.

The machine will change the internal state of all its processors at given instants of time by means of the application of F; the evolution of the machine may now be described in terms of the outputs of the PEs, namely the $O^t_{xy} = (d,e)_{xy}$ where "d" represents the output binary value of the new pixel whilst the second variable "e" is the value of the output signal which is sent to all the specified neighboring elements (up to a maximum of 8).

Within this framework, the CLO computes the output state (O^t_{xy}) for each PE of the array having coordinates x,y from its internal state (I^t_{xy}); such a CLO is simultaneously and independently computed on all the pixels of the image. In conclusion, the parallel processing of information at time t may be represented by a sequence of the output states of all the processors $(O^0_{xy}, O^1_{xy}, O^2_{xy}, \ldots O^t_{xy})$.

As described in more detail than here in (12), the sequences of CLOs turn to be the tools for programming a parallel machine to perform some useful task. Such sequences may be artificially subdivided into three classes: 1) the unit sequence of a single CLO; 2) the same CLO repeated a number of times and 3) a sequence of different CLOs which might itself be repeated a number of times.

XII. PRACTICAL CLOs

Let us now give some examples that will show the different classes of CLOs in practice. The boolean operation (like AND, OR, etc.) between two images (i.e. between their corresponding pixels) corresponds to a CLO that belongs to the first class. Similarly, the extraction of a contour from a binary image on the basis of the presence of a 0-element in the 8-neighbourhood of a pixel also represents a class 1) CLO.

Whenever the components of an image must be shifted, for instance towards the right, we may iterate a CLO that will preserve or turn an element into a 1-element provided its left neighbour is a 1-element. The repetition of such a CLO constitutes an example of a class 2) sequence and will either require a fixed number of iterations (established a priori) or a testing mechanism in order to stop the right-shifting CLO.

Finally, a CLO sequence of class 3) is the one required for object extraction in our image which has many objects and a background and every single object must be processed independently, (see for instance (30)). This sequence will be used for the extraction of the first object (by means of a scanner) then again for the second object and so on until all the objects present in the image have been processed.

Typical CLOs are, for instance, shrink (expand) which enables a contraction (growth) of an object (according to the chosen neighborhood) at each iteration of the CLO. Such a reduction is often called core extraction and one well-known smoothing algorithm used in preprocessing images uses a combination of shrink - expand CLOs which eliminates little notches and bays on the contour of an object. Another interesting application of shrinking is to count the number of 1-elements which remain after each shrinking step so generating a size histogram (31); by plotting this number against the step number; large objects will remain until the last shrinking step. Finally, shrinking might also be used for counting connected objects (32) on an image and after each shrinking step only the isolated 1-elements will be counted and deleted until no more elements remain. This process must obviously avoid splitting a connected component and merging two close components on the image.

There are also other well known image transformations which might be used both for preprocessing or for shape description (for an overall description of these techniques see (5)) and they all fall into CLOs of class 3). Thinning, skeletonizing, templating and matched filtering are some of the most usual ones and the literature provides a vast repertoire of algorithms most of which may be expressed by a sequence of CLOs. For example the applications of deletion rules to 1-elements to erase those elements which will not disconnect the object is equivalent to the computation of a CLO on a three by three neighborhood in which the central element might be turned to 0 if some conditions hold on its neighbors and this basic procedure is iterated for all the eight possible directions a number of times until the output image does not change any more: this is the termination test for a data-dependent CLO string.

From a slightly different angle we might also consider local digital filtering as a class 3) CLO string in which neighborhoods (not necessarily square but more generally rectangular) are established together with the values of weights in order to extract contours in grey level images. The difference filter (33), for instance, is defined as follows:

$$D_1[L1, L2 ; K1, K2] : I \rightarrow O \qquad O^t_{xy} = I^t_{x+L2,\, y+K2} - I^t_{x-L1,\, y-K1}$$

and

$$D_2[L1,L2;K1,K2] : I \rightarrow O \quad O^t_{xy} = I^t_{x+L2,y+K2} - 2I^t_{xy} + I^t_{x-L1,y-K1}$$

Such filters may be applied to smoothed images for reducing the effect of random noise and, therefore, as a second part within the preprocessing stage.

In practice for a border detection, a difference filter may be used and then thresholding and thinning may follow so as to obtain a one element thick digital contour from a grey level input image.

XIII. CONCLUSIONS

We have tried to give some flavour of the different interpretations of the word "parallel" from the point of view of the computer word, computer arithmetic, computer architecture, local computation, high level language and basic sequence of operations to perform an image processing task. From a disordered sequence of approaches which were dependent both on the author's background and on the available technology of the times, a clearer picture may be seen to grow up if one considers that the key point in obtaining computational speed up is to well match the hardware architecture of the system with the algorithm that will compute a function on the input image. Furthermore, if a language exists (34) for expressing this function in terms of a local operation which can be performed simultaneously and independently on all the pixels of the image, this will enable us to write programs in a way which is both clear for the programmer and directly executable by an array processor. This important point was recognized and discussed also in connection with sequential progressing in the well known (35). This goal appears now to be reachable within the next few years, capitalizing on the experience that has been gained in all the work done until now even if differently motivated.

REFERENCES

1. "Pattern Recognition", The Journal of the Pattern Recognition Society, ed. R.S. Ledley, Pergamon Press, U.K.

2. "Computer Graphics and Image Processing", eds. A.Rosenfeld M. Freeman, T.S. Huang, A. von Dam, Academic Press, USA, U.K.

3. IEEE Transaction on Pattern Analysis and Machine Intelligence, ed. K.S.Fu, IEEE Conputer Society, USA.

4. H.Kazmierczak, "Processing of image data for remote sensing application" Proc. NATO-ASI Digital Image Processing and Analysis, Eds. J;C. Simon, A.Rosenfeld, pp.317-338.

5. A. Rosenfeld, A.C. Kak, "Digital Picture Processing", Academic Press, 1976.

6. Pavlidis, "Structural Pattern Recognition", Springer Verlag, 1977.

7. S.M. Selkow, "One-pass complexity of digital picture properties", Journal of ACM, 19, 283, 1972.

8. L. Uhr, "Layered Recognition Cone Networks that Preprocess, Classify and Describe", IEEE Trans. Comp.21, 758-768,1972.

9. M.D. Levine, J. Leemet, "A method for non-purposive picture segmentation", 3rd IJCPR, Coronado, California, 494-498, 1976.

10. L.S. Davis, "A survey of edge detection techniques", CGIP, vol. 4, 3, 248-270, 1975.

11. S.W. Zucker, "Region growing: childhood and adolescence", CGIP, vol. 5, 3, 382-399, 1976.

12. J. Weszka, R.N. Nagel, A. Rosenfeld, "Steps toward handwritten signature verification", 1st IJCPR, Washington, 59-66, 1973.

13. L. Cordella, M.J.B. Duff, S. Levialdi, "Thresholding: A Challenge for parallel processing", CGIP, vol. 6, 3, 207-220, 1977.

14. S.L. Horowitz, T. Pavlidis, "Picture segmentation by a directed split-and-merge procedure", 2nd IJCPR, Copenhagen, 424-433, 1974.

15. M. Nagao, T. Matsuyama, "Focus of attention in the analysis of complex pictures such as aerial photographs", Japan United States Seminar Research Towards Real-Time, Parallel Image Analysis and Recognition (Oct. 31-Nov. 4, 1978), Tokyo, Japan.

16. M. Nagao, T. Matsuyama, "Edge preserving smoothing", 4th IJCPR, Kyoto, 518-520, 1978.

17. M. Nagao, "Region extraction and shape analysis of aerial photographs" 4th IJCPR, Kyoto, 620-628, 1978.

18. J. Sklansky, Pattern Recognition, 2, 3-10, 1970.

19. T. Pavlidis, "Algorithms for shape analysis of contours and waveforms" 4th, IJCPR, Kyoto, 70-85, 1978.

20. M.J. Flynn, "Some Computer Organization and their Effectiveness" IEEE Trans. Comp. C-21, 948-960, 1970.

21. D.J. Kuck, "A Survey of Parallel Machine Organization and Programming", Computing Surveys, vol. 9, no.1, 29-59, 1977.

22. S.S. Yau, "Associative Processor Architecture - A Survey", Computing Surveys, vol. 9, no.1, 3-27, 1977.

23. S.Levialdi, A. Maggiolo-Schettini, M. Napoli, G. Uccella, "Consideration on parallel machines and their languages" Workshop on High Level Languages for Image Processing, Windsor, June 1979.

24. S. M. Unger, "A Computer Oriented towards Spatial Problems", Proc. IRE, 1744-1750, 1973.

25. M.J.Duff, "A Cellular Logic Array for Image Processing", Pat. Recog. vol. 5, 229-247, 1973.

26. B. Kruse, "A Parallel Picture Processing Machine, IEEE Trans. on Comp. C-22, n. 12, 1075-1087, 1973.

27. L.P. Cordella, M.J.B. Duff, S. Levialdi, "An analysis of Computational Cost in Image Processing: A Case Study", IEEE Trans. on Comp. vol. C-27, no.10, 904-910, 1978.

28. L. P. Cordella, M.J.B. Duff, S. Levialdi, "Thresholding: a Challenge for Parallel Processing", Comput. Graphics and Image Processing, vol. 6, 207-220, 1977.

29. K. Preston K. Jr., M.J.B. Duff, S. Levialdi, P.E. Norgren, J. Toriwaki, "Basics of Cellular Logic with Some Applications in Medical Image Processing", Proc.IEEE, 826-856, May 1979.

30. C. Arcelli, L.P. Cordella, S. Levialdi, "Parallel Processing for Image Analysis", NATO ASI on New Concepts and Technologies in Parallel Information Processing, ed. E.R. Caianiello, Nordhoff Int. Pub., Leyden, 105-121, 1975.

31. M. Ingram, K. Preston Jr., "Automatic Analysis of Blood Cells", Sci. Amer.223 (5), 72-82, 1970.

32. S. Levialdi, "On Shrinking of Binary Patterns", Comm. ACM 15(1), 7-10, 1972.

33. S. Yokoi, J-i, Toriwaki, T.Fukumara, "Theoretical Analysis of Parallel Processing of Pictures Using Algebraic Properties of Picture Operators", Proc. 3rd IJCPR, 723-727, 1976.

34. S. Levialdi, A. Maggiolo-Schettini, M. Napoli, G. Uccella, "PIXAL: a High Level Language for Image Processing, working Paper", Japan United States Seminar, Research towards realtime parallel image analysis and recognition, Tokyo, 1978.

35. A. Rosenfeld, Pfaltz I.L;, "Sequential Operations in Digital Picture Processing", Journal ACM, vol. 13, n.4, 471-494,1966.

MAP SEQUENCE PROCESSING*

G.G.Pieroni

M.F.Costabile

C.Guerra

GNAFA-CNR

Dipartimemto di Matematica

Università della Calabria

Roges di Rende, Cosenza, Italy

I. INTRODUCTION

Automatic analysis of image sequences is becoming increasingly important in many sectors of applied science. In some particular areas, as for example biomedical applications and territorial monitoring, it would appear that automatic analysis of dynamic documents now plays a determining role in the interpretation of the phenomema expressed by such documents. This problem has recently been of considerable interest to various authors (1).

Potter (11) uses motion information to segment a picture.

*This research has been partially supported by NATO Grant 1261.

ISBN 0-12-267180-5

Points which have the same velocity are grouped together to form a basic object. The velocity of a point is calculated by defining a cross template around it and by searching for a matching template in the successive frame. The displacement between points is a measure of the velocity. In (12) an approach for the analysis of the motion of moving granules in a pancreatic cell is described. A picture giving the areas of change between two images is made simply by subtracting one picture from another. Positive and negative parts (called backward and forward shadows) are extracted from the picture by thresholding. The pairing of backward and forward shadows corresponding to true motion is made on the basis of certain features of each segment such as area, centroid, direction of the principal axes, etc. In (13) a heuristic method for tracing the boundary of the left ventricle from cineangiographic images is proposed. In detecting a boundary in a frame, the information contained in the previous frame is also utilized. The boundary tracing algorithm is as follows: compute the spatial derivatives of the gray levels so as to get the gradient image. Add to each point of the gradient image a weight coefficient according to the previous frame. Then draw a continuous curve by joining points with the local maximum values on the gradient image;(the motion will be considered the change in the distance between the centroid and a point selected on the boundary). Endlich et al.(14) deal with automatic analysis of cloud motion from satellite photographs. Thresholding and clustering techniques are used to extract the clouds from their background. Then their centroids, called 'brightness centers', are calculated. The matching between centers in two contiguous frames is based on an evaluation of brightness and distances between centers. This process yields a different motion measurement for each pair of centers. In (15) a system that detects and tracks moving objects from natural processes recorded on videotape or cine film is described. The algorithm for extracting objects from the background is based on spatial and temporal differences of gray levels. The problem of occlusion is also considered. To track objects between scenes, models of the previous frames and a prediction map are all used. The prediction map estimates whether an object is occluded or not, and if so what object occludes it. In case of an uncorrect prediction, a backtracking process reanalizes those objects misinterpreted in the tracking process. Aggarwal and Duda (16) present an idealized model of the problem of analyzing pictures. The scenes are not real scenes, but are artificially generated containing rigid polygonal objects. The problem is to determine the linear and angular velocities of the objects. The occlusion problem is emphasized. Vertices of an image are classified as 'real' and 'false' vertices. 'Real' vertices are actual vertices of the polygons; 'false' vertices are vertices generated by the intersection of occluding polygons. The angular measure of the 'false' vertices from one picture to another will change; whereas the angular measure of 'real' vertices will not change. In (17) the basic problem is to track the motion of planar curvilinear objects. The linear and angular velocities of the objects are computed and a model is created and updated at each scene. A prediction of the next scene is also generated from the estimated velocities.

From the above three principal tendencies emerge:

1) Given a frame I_i, belonging to a sequence $I=I_1,I_2,\ldots,I_n$, information contained in contiguous frames $(\ldots I_{i-2},I_{i-1},$

$I_{i+1}, I_{i+2}, \ldots$) is used in an attempt to consolidate certain choises in the segmentation process of I_i (5-6),(11-13),(15).

2) Given a frame sequence I, the boundary of an object consid-dered rigid is determined on a frame $I_i \in I$. By comparing this boundary on successive frames $(I_{i+1}, I_{i+2}, \ldots, I_{i+m})$ with respect to a fixed reference system, they try to extract motion parameters for the object in the image domain (rotation-translation). This procedure could, where neces-sary, take into account a possible occlusion of parts (11), (12),(15-17).

3) Let P be considered a dynamic process, it often makes sense to represent it by means of a frame sequence where one or more elements, generally represented by shape boundaries, gradually evolve by a modification of the structure of the line representing the boundary in question (2),(5),(6),(13).

In the latter case a possible method for analyzing the se-quence can be conceived by defining a sequence of transfer op-erators $T_{1,2}, T_{2,3}, \ldots, T_{n-1,n}$ each of which (e.g. $T_{i-1,i}$) is able to give the frame I_i when applied to I_{i-1} ($T_{i-1,i}(I_{i-1}) = I_i$). Various formulations can be suggested for the operator T. Considering however the quantity of data involved and the time required for processing them, it would seem clear that such operators must be characterized by the following three fundamental properties:

a) They should be easily formulated in order to save computing time in the trasformation operations.

b) It should be possible to define inverse transformation op-erators in order to reach the same results when scanning the sequence in the opposite direction. In other words, given an operator $T_{i-1,i}$ such that $I_i = T_{i-1,i}(I_{i-1})$ it should be possible to determine an operator $\overline{T}_{i,i-1}$ such that $I_{i-1} = \overline{T}_{i,i-1}(I_i)$.

c) They should all be expressible in terms of structure of a limited number of elementary operators, $T\ (t_o,t_1,\ldots,t_n)$, where $t_o,t_1,\ldots,t_n$ are elementary basic operators given a priori for the entire image sequence.

Considering the difficulty of planning simple operators capable of mapping an image I_{i-1} on I_i it is worth trying a mapping operation on the structure rather than on the images themselves. Formally, the process can be expressed very simply in the following terms. Given a sequence $I=I_1,I_2,\ldots,I_n$ the corrisponding one obtained by extracting the structure of each frame is considered:

$$sI = sI_1,sI_2,\ldots,sI_n$$

In this context it would make sense to determine an operator sequence $T^s_{1,2},T^s_{2,3},\ldots,T^s_{n-1,n}$ such that

$$sI_i = T^s_{i-1,i}(sI_{i-1})$$

which would verify the properties expressed previously. The sequence $T^s_{1,2},T^s_{2,3},\ldots,T^s_{n-1,n}$ is called 'the dynamic structure' of the image sequence. Extreme care must, however, be taken in the application of such a theory since the information compression in the passage image → structure makes the interpretative model of each sequence extremely rigid. As a consequence an uncorrect evaluation of a frame, (e.g. in the segmentation process) could lead to the identification of structural transfer operators which are meaningless in the process in question. In effect the segmentation of each single frame constitutes a crucial point in the cognitive process of a frame sequence.

II. DYNAMIC SEGMENTATION

The concept of using information extracted from contiguous frames in order to segment a given frame correctly has already been considered by some authors (5-6),(11-13),(15).

Our approach is based essentially on the following consideration. It is possible in many cases to determine 'unprecisely' boundary of parts inside an image using techniques already known (4). This lack of precision is due to the fact that a clear boundary cannot always be detected but rather a 'band' within which the required boundary falls. Supposing that the sequence I has been transformed into another sequence C, composed for every frame of a suitable boundary band C_i extracted from I_i using known methods, it would make sense to reinforce the definition of the real part of the boundary on the band C_i using information extracted from contiguous frames (Fig.1a,1b). We will thus assume, as a working hypothesis, that every modification in the boundary of the object to be segmented corresponds to a modification in the respective band C_i while modifications occurring in the band are insufficient to confirm a modification in the boundary of the object. In other words, considering a particular variation in the band from C_{i-1} to C_i , we will assume that such a variation does not in itself guarantee the presence of a modification in the object boundary, since it could be due to noise or other factors connected to the digitization of the image. In reality the boundary extraction from the band could only be accomplished if a certain modification in the band itself results from a real movement of the object to be segmented.

The consolidation of the movements is based on the following algorithm:

Let two contiguous bands be C_{i-1} and C_i , the evaluation of their difference pixel by pixel leads to the extraction of a multitude of parts candidates to be variation zones in the passage from C_{i-1} to C_i and which will be called 'islands' (5-6). Only some of the islands, appearing when two bands are superimposed, represent the trace of a real movement performed by the shape in the passage from C_{i-1} to C_i. In order to identify them we shall limit our research to variations which have the following properties:

a) A flow structure.

b) Consistency in a minimum time interval (minimum number of frames).

If we consider the superimposition of the two bands C_{i-1} and C_i (Fig.1c), we can divide the set obtained in this way into the following three zones:

$$C_i \cap C_{i-1} = D(C_i,C_{i-1})$$

$$C_i - C_{i-1} = E(C_i,C_{i-1})$$

$$C_{i-1} - C_i = C(C_i,C_{i-1})$$

The zones $D(C_i,C_{i-1})$, $E(C_i,C_{i-1})$ and $C(C_i,C_{i-1})$ shall be respectively called the 'intersection', 'expansion' and 'contraction' zones. We shall say that there is a flow structure in a variation zone between the two bands C_{i-1} and C_i when it is possible to determine one or more expansion zones associated with one or more contraction zones (5-6). These zones are determined as follows:

1) The islands having a maximum diameter less than a given threshold are eliminated (very small islands are considered noise).

2) The islands are subdivided into groups having similar 'trim' in a reference system O(xy).

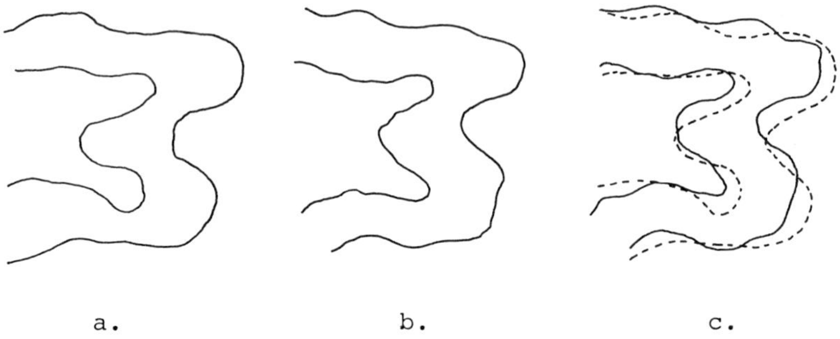

FIGURE 1.

By considering the distance between the centroids of E and C-islands identified in this way, we say that an E-island is associated to a C-island when this distance is not greater than the average of the maximum diameters of the two islands (Fig.2).

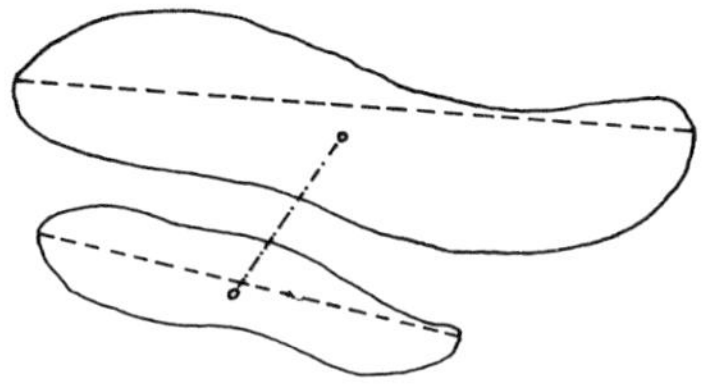

FIGURE 2.

For every group of associated islands we will now consider the respective E-islands and will compute the max angle M (flow angle), $|M|<\pi$, obtained by joining the centroids of each E-island with the centroids of the C-islands associated with it (Fig.3). We shall assume the bisecting line of the M angle to be the flow direction (sometimes only one C-island is connected to an E-island; in this case the flow is assumed to be the line joining the centroids of the two islands in the direction C to E-island).

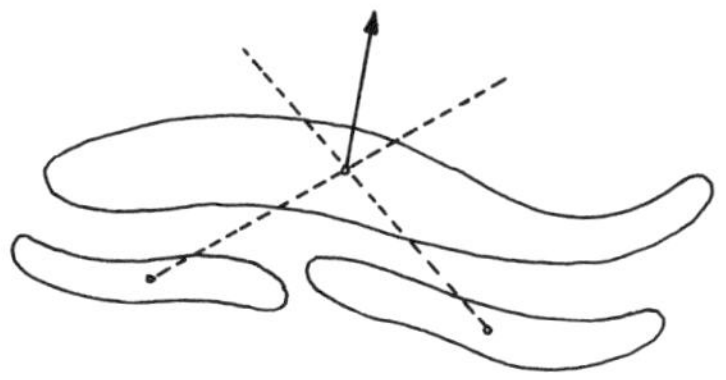

FIGURE 3.

It is convenient to assume that the intensity of the flow vector is equal to the normalized average of the surfaces of the E and C-islands connected in that flow. The point of application is given by the centroid of the E-island. This procedure leads to a sequence of flow maps for each couple of frames C_{i-1}, C_i as represented in Fig.4.

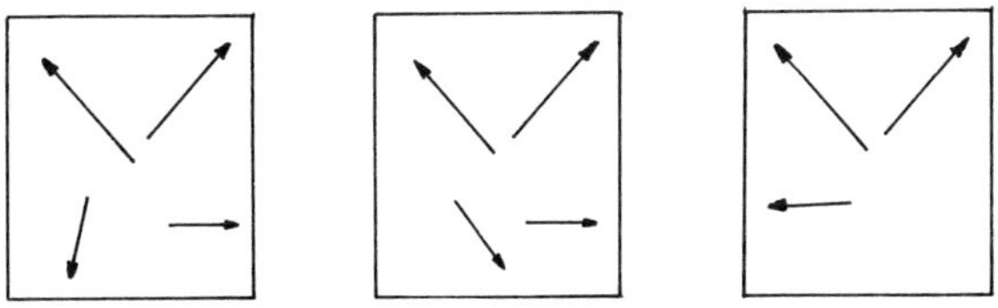

FIGURE 4.

The persistence of a given flow in a certain zone of the frame C_i for a given number of successive frames (considered a parametric number) reinforces the hypotesis of the real consisstency of that movement (5-6).

By considering the band represented in Fig.1.b, where a consolidation of the movement was accomplished, it can be clearly assumed that the two protrusions and the central intrusion really belong to the structure of the image and have not been generated by noise. Thus it is possible to obtain a boundary line for the image C_{i-1} just by extracting the external boundary of the band which is there represented.

III. DYNAMIC STRUCTURE

The analysis of a sequence of images implies the formulation of a model capable of describing the sequence itself. In reality a cascade sequence of operators seems to constitute a feasible descriptive model. The mechanism is based on the fundamental fact that once the operator sequence has been extracted from a comparison between contiguous images it is possible, given the first frame of the sequence, to reconstruct the entire information by transforming initial signals.

Thus when the operator sequence is obtained it is clear that it does, in effect, represent a pattern that can be used for the purpose of automatic recognition in a sequence universe. It has already been pointed out how complex it is to define simple transfer operators directly onto the images. Given the importance of the structure in the cognitive process for this kind of problem, the use of an analysis model of the structure rather than a direct model on the images does not appear to constitute a serious limitation. Thus the concept 'dynamic structure of an image sequence' is carried out representing a sequence of operators which transform the structure along the sequence evolution. The crucial point in this approach consists in the formulation of the transfer operators. A possible extraction procedure for the dynamic structure is, given a sequence of shape boundaries, as follows:

a) The definition of a grammatical scheme in order to parse each shape boundary.

b) The definition of a set of elementary transfer operators.

c) The decomposition of each single shape boundary and the extraction of its structure.

d) The evaluation of structural differences expressed in terms of elementary transfer operators.

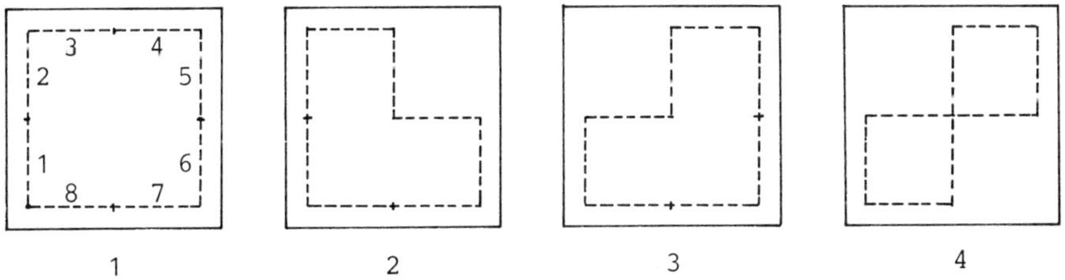

FIGURE 5.

So as to clarify the concepts previously expressed we will develop the following simple example. Given the sequence expressed in Fig.5, composed of four frames, the grammatical scheme given by the elements represented in Fig.6 will be considered (9).

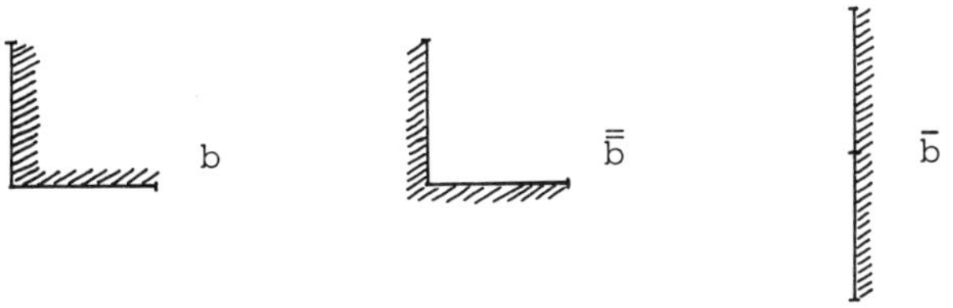

FIGURE 6.

The sequence can now be written as in scheme 1.a .

		1		2		3		4
								Frame Number
	1	b		b		b		b
Segment	2	$\bar{b}$		$\bar{b}$	→	b		b
Number	3	b		b	→	$\bar{\bar{b}}$		$\bar{\bar{b}}$
	4	$\bar{b}$	→	b		b		b
	5	b	→	$\bar{\bar{b}}$	→	b		b
	6	$\bar{b}$	→	b	→	$\bar{b}$	→	b
	7	b		b		b	→	$\bar{\bar{b}}$
	8	$\bar{b}$		$\bar{b}$		$\bar{b}$	→	b

SCHEME 1.a

By introducing the following basic elementary transfer operators

$$t = b \to \bar{b}, \quad \bar{t} = \bar{b} \to b, \quad \hat{t} = b \to \bar{\bar{b}}, \quad \hat{\hat{t}} = \bar{\bar{b}} \to b, \quad \check{t} = \bar{b} \to \bar{\bar{b}}, \quad \check{\check{t}} = \bar{\bar{b}} \to b$$

we obtain the dynamic structure expressed in scheme 1.b.

	1→2	2→3	3→4
1			
2		$\bar{t}$	
3		$\hat{t}$	
4	$\bar{t}$		
5	$\hat{t}$	$\hat{\hat{t}}$	
6	$\bar{t}$	t	$\bar{t}$
7			$\hat{t}$
8			$\bar{t}$

SCHEME 1.b

It is obvious from the result obtained in this example that it is possible to group together the operators in a hierarchy. Thus the dinamic structure of the sequence given in the example will be transformed as represented in scheme 2, which expresses the descriptive model of the modifications successively intervening in the initial configuration. In fact, by introducing the structure $b\,\bar{b}\,b\,\bar{b}\,b\,\bar{b}\,b\,\bar{b}$ into the initial operator $\mathring{t}_{4,5,6}$,the result obtained is the structure $b\,\bar{b}\,b\,b\,\bar{\bar{b}}\,b\,b\,\bar{b}$ and so on in a cascade.

$$
\begin{array}{lll}
1 : \mathring{t}_{4,5,6} & & \mathring{t} = \bar{t}\,\hat{t}\,\bar{t} \\
2 : \overset{-\circ-}{t}_{2,3} \ ; \ \overset{+}{t}_{5,6} & \text{where} & \overset{-\circ-}{t} = \bar{t}\,\hat{t} \\
3 : \mathring{t}_{6,7,8} & & \overset{+}{t} = \hat{\hat{t}}\,t
\end{array}
$$

SCHEME 2.

Since the modification occurring between frames must, by definition,be small it is clear that the dynamic structure must, in normal cases, be very simple. It is also clear that both the grammatical scheme and the relative shape decomposition prove very elementary in the example given above. This approach can be expressed in more general terms by considering a sequence of artificially generated contour lines. Let us give a sequence of shape boundaries $M=M_1,M_2,\ldots,M_n$ expressed by closed polygonal lines free from noise. Let us also suppose that the modifications occurring in a line M_i in the passage from M_{i-1} to M_i always concern one uninterrupted sector of the line only. In other words we will assume that modifications occurring separately and contemporaneously on the line are impossible. Moreover, in this work we will not treat modifications obtained by means of rigid rotations and translations of the entire shape. We will model the condition previously described in the following terms. Let us consider a Cartesian reference O(xy) where n points $P_1,P_2,\ldots,P_n$ are the vertices of a regular polygon; by using spline functions for interpolating (10), a parametric number of auxiliary points p_i will be introduced between the basic points P_i and P_{i+1}, $i=1,2,\ldots,n$ and $P_{n+1}=P_1$ (Fig.7).

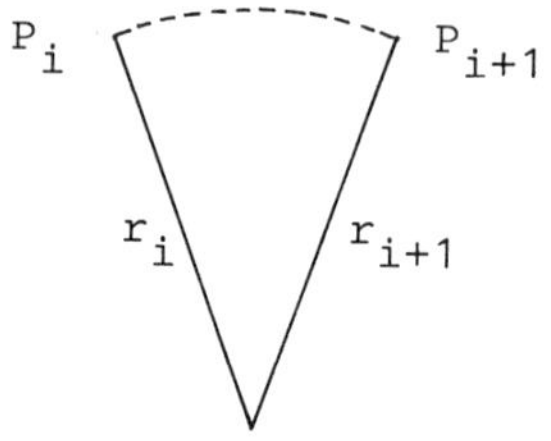

FIGURE 7.

It is supposed that the distance $d(p_i,p_{i+1})$ is pratically constant. The decomposition obtained considering the basic points as separating points, identifies elements of the polygonal line which we will indicate as basic segments. Now we will consider the following grammatical scheme with attributes (scheme 3).

Code	Segment	Segment Length	Chord Length
a	Concave arch	l_o	d_o
b	Convex arch	l_1	d_1
c	Line segment	l_2	d_2
d	Inflexion	l_3	d_3

SCHEME 3.

By using such a scheme it is possible to obtain the decomposition of a line M of the sequence in terms of basic segments with attributes.

The hypotesis of a 'slow variation' of the shapes between two following frames is realized by modifying k contiguous segments, $k << n$ (n is the number of basic segments). With this hypotesis it is possible to construct a transition table formed by the elementary transfer operators, applicable to the single elements of the line on the basis of a comparison between the decomposition of M_{i-1} and M_i (see scheme 4).

1	t_0	Neutral: a→a, b→b, c→c, d→d.
2	t_1	Exchange one step on scheme 3 (e.g. a→b).
3	t_2	Exchange two steps on scheme 3 (e.g. a→c).
4	t_3	Exchange three steps on scheme 3 (e.g. a→d).
5	t_4	Small positive variation in the curve-chord ratio.
6	t_5	Big positive variation in the curve-chord ratio.
7	t_6	Small negative variation in the curve-chord ratio.
8	t_7	Big negative variation in the curve-chord ratio.
9	t_8	Small positive variation in the curve length.
10	t_9	Big positive variation in the curve length.
11	t_{10}	Small negative variation in the curve length.
12	t_{11}	Big negative variation in the curve length.

SCHEME 4.

Let us generate the artificial sequence by following the steps below:

a) The original configuration M_1 is obtained by interpolating among the points $P_1, P_2, \ldots, P_{15}$ (10).

b) By extracting a random number n_1, $n_1 \in \{1, 2, \ldots, 15\}$, three radii r_{n_1-1}, r_{n_1}, r_{n_1+1} are considered (Fig.7) (r_i = distance between the polygon center and the P_i point).

c) By extracting a second random number n_2, $n_2 \in \{0,1\}$, the three radii are lengthened or shortened by a predetermined quantity.

d) By interpolating again (10) the configuration M_2 is obtained and so on for a sequence of frames.

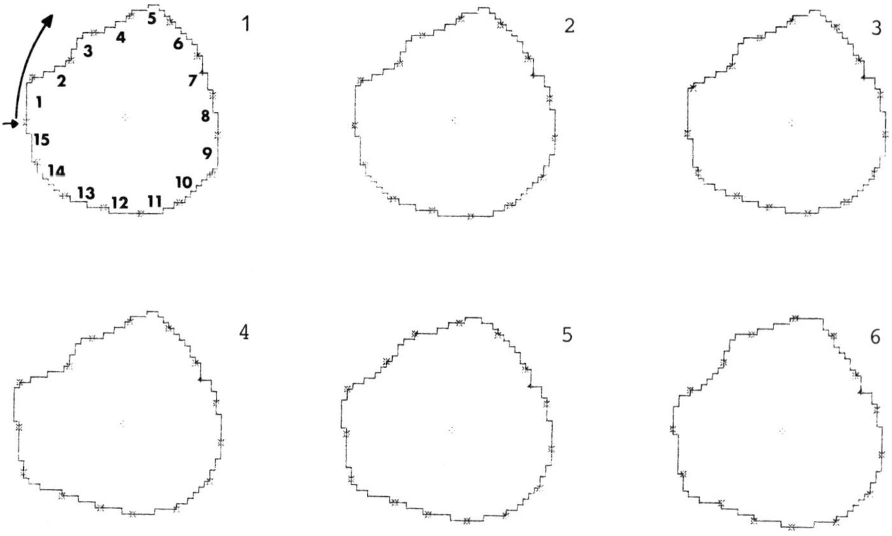

FIGURE 8.

Fig. 8 shows frames belonging to the sequence obtained by starting from a given configuration. It is evident that the sequence obtained in this way has properties previously expressed. The application of scheme 3 leads to the extraction of the structure for each frame as seen in scheme 1.a. The use of the operators given in scheme 4 leads to the formulation of the dynamic structure as in schemata 5. It is obvious from scheme 5.b that, given the initial geometrical structure, the operator

Segment Number	1	2	3	4	5	6
1	b	b	b →	d	d →	b
2	a	a	a	a →	a →	a
3	b	b	b	b →	b →	b
4	a	a	a	a →	a	a
5	b	b	b	b →	b	b
6	b →	b	b	b	b	b
7	d →	d	d	d	d	d
8	b →	c	c	c	c	c
9	b →	b	b	b	b	b
10	c	c	c	c	c	c
11	b	b →	b	b	b	b
12	b	b →	d	d	d	d
13	a	a →	d →	a	a	a
14	b	b →	b →	b	b	b
15	a	a	a →	b	b →	a

Frame Number (columns 1–6)

SCHEME 5.a

	1→2	2→3	3→4	4→5	5→6
1	t_o	t_o	t_2	t_o	t_2
2	t_o	t_o	t_o	t_4	t_6
3	t_o	t_o	t_o	t_6	t_5
4	t_o	t_o	t_o	t_8	t_o
5	t_o	t_o	t_o	t_6	t_o
6	t_o	t_o	t_o	t_o	t_o
7	t_8	t_o	t_o	t_o	t_o
8	t_1	t_o	t_o	t_o	t_o
9	t_4	t_o	t_o	t_o	t_o
10	t_o	t_o	t_o	t_o	t_o
11	t_o	t_6	t_o	t_o	t_o
12	t_o	t_2	t_o	t_o	t_o
13	t_o	t_3	t_1	t_o	t_o
14	t_o	t_4	t_o	t_o	t_o
15	t_o	t_o	t_1	t_o	t_3

SCHEME 5.b

sequence can reconstruct the evolution of the shapes. In order to put scheme 5.b to a practical purpose it would be greatly desiderable to have a method for a hierarchical composition of the given operators. This problem is now the subject of several studies.

IV. CONCLUDING NOTES

The problem of map sequence analysis has not yet been given the attention it deserves in spite of its many practical applications. In section I we have tried to underline the fact that such applications are not restricted to the determination of the cinematics of rigid elements only. In fact, if we consider a sequence composed of single rigid elements, moving in a reference system, the problem of analyzing the sequence can be reduced to that of identifying each component element and then tracking its cinematic parameters on the time axis. The problem can however be complicated by adding conditions concerning the occlusion of objects which would obviously introduce factors difficult both to evaluate and interpret. The problem which we intend to pose regards methods for analyzing sequences of shape boundaries whose structure modifies in the course of time. There are, however, some questions to be asked, each of which represents a sector in which an exhaustive answer would be extremely desirable. The main questions appear to be the following:

i. How to make a segmentation in order to obtain related segments between a frame and the following ones (normalized segmentation).

ii. How to parse a shape boundary taking into account the

behaviour of the related shapes in order to obtain a normalized parsing over the sequence.

iii. How to infer a grammar valid for every frame.

iv. How to infer a dynamic grammar (transition grammar) and what relationship there is between the dynamic grammar and the previous one.

v. How the relational rules can be transformed from a two-dimensional to a three-dimensional context.

In this work a number of problems connected with these questions has been raised but proposals for their solution are, at present, only in the elaboration phase.

REFERENCES

1. Martin, W.N., and Aggarwal, J.K., *C.G.I.P. 7*, 356 (1978).
2. Pieroni, G.G., *C.G.I.P. 10*, 375 (1979).
3. Pieroni, G.G., and Freeman, H., *4IJCPR*, 731 Kyoto, Japan, November 1978.
4. Weszka, J.S., and Rosenfeld, A., Computer Science Technical Report TR-349 University of Maryland, 1975.
5. Pieroni, G.G., and Costabile, M.F., *IEEE Computer Society Conference on PRIP*, 300 Chicago, USA, August 1979.
6. Costabile, M.F., *C.G.I.P.*, in press.
7. Shapiro, L.G., and Haralick, R.M., *IEEE Trans. on PAMI*, 1, 10 (1979).
8. Guerra, C., and Pieroni, G.G., Istituto di Automatica Technical Report TR 80-02 Università di Roma, 1980.
9. Bibriesca, E., and Guzman, A., Comunicaciones Technicas from Instituto de Investigaciones en Matematicas Aplicadas y en Sistemas, Universidad Nacional Autonoma de Mexico, TR 178, 1978.
10. Pieroni, G.G., *Calcolo 5*, 173 (1968).
11. Potter, J., *C.G.I.P. 6*, 558 (1977).
12. Takagi, M., and Sakaue, K., *4IJCPR*, 735 Kyoto, Japan, November 1978.
13. Eiho, S., and Kuwahara, M., *4IJCPR*, 740 Kyoto, Japan, November 1978.
14. Endlich, R.M., et al., *J. of Appl. Meteorol. 10*, 105 (1971).
15. Yachida, M., Asada, M., and Tsuji, S., *4IJCPR*, 726 Kyoto, Japan, November 1978.
16. Aggarwal, J.K., and Duda, R.O., *IEEE Trans. on Computers C24*, 966 (1975).

17. Chow, W.K., and Aggarwal, J.K., *IEEE Trans. on Computers C26*, 179 (1977).

18. Freeman, H., *Computing Surveys*, 6, (1), 1974.

NUMERICAL ALGORITHMS FOR INTERPOLATION AND SMOOTHING

Helmut Werner

Institut für Numerische und instrumentelle Mathematik
und Rechenzentrum
Westfälische Wilhelms-Universität Münster/Westfalen

This paper is an abbreviated version of lectures on the principles and methods of approximating functions of one and two variables. Its intention is tutorial, therefore publications of elementary level as well as advanced and research level are quoted. The first part deals with the concepts developed for functions of one variable in order to illuminate the analogous trend for two variables described in the second part. (Most methods would easily carry over to higher dimensions.) The quoted simple approach of Newton type interpolation for two variables given in (26) and two dimensional histosplines (9) seems to be so far unobserved in the literature. A complete text of the given lecture including many graphs and examples illustrating these techniques is found in (25).

1. INTRODUCTION

1.1 In this lecture I will try to present a summary of techniques and programs for the documentation and representation of functions of one or two real variables, i.e. for curves and surfaces that have been made available by "Applied mathematicians" within the last years to support the scientists in the applied field in their work. It is obviously impossible to give

ISBN 0-12-267180-5

anything like an exhaustive survey of the literature pertaining to this field. I have therefore tried to quote books and articles that give methods that have already been turned into codes and that will give references to those sources which finally describe codes concerning the representation and smoothing of data.

1.2 To give a precise answer the mathematician needs a clearly and precisely stated question. However in applications we frequently meet a rather hazy statement about what should be achieved mathematically. It is not so frequent that we are in the comfortable situation to have a <u>well posed problem</u> that is (rather) accurate data and a set of equations, which allow for a clear answer. A well posed problem has one and only one solution and this is stable with respect to errors (or noise) of the data, this means that small changes of the data cause only small changes of the answer. Typically well posed problems can be formulated in physics.

1.3 If we turn to biology or medicine the data are usually very poorly defined. In fact there are cases in which there are only qualitative measurements like an evaluation on the scale "very good, good, fair, bad". In this case one may try to classify the objects, characterized by a number of properties and/or variables with respect to certain criteria and split them into groups of similar objects. These groups are sometimes called clusters. We will not enter this field but refer the interested reader to the books by Späth (21) and Steinhausen/Langer (22) which try to describe the <u>cluster analysis</u> with minimal requirements in mathematics and give programs and worked examples.

1.4 In this talk we infer as a usually not explicitely stated assumption that the data to be represented are primarily structured due to a law of nature and hence their "true values" bear a certain degree of regularity. The gathered data may be falsified by some noise from the measurement or smaller effects left out of consideration.

During this conference I learnt that this assumption may be heavily violated in case of geographical data which might be of highly accidental nature due to the many different natural phemomena and the interference of man.

If the assumption on smoothness seems justified it is wise to look for analytic models first. Its solutions or approximations of solutions can usually be specified by a small number of parameters. These parameters ought to be chosen so that the measured data are fitted. This approach allows an extremely high condensation of the obtained information with respect to storage and evaluation. This type of tailor-made solution should be preferred if it is applicable, compare (1).

In case a model can be set up successfully there is still the need of transforming it into an operational algorithm. First one has to find an appropriate numerical method. Next one has to supervise the error inferred by discretization, if limits were involved. Furthermore the accumulation of round off errors has to be monitored during the application to data. Keeping the later ones small may cause an implementation on a computer that looks efficient because of its thrifty number of arithmetic operations ("complexity") to be replaced by another one that takes more time but is "well conditioned" with respect to the influence and accumulation of round off errors.

If forming a specific model is not feasible for one or the other reason general interpolation or approximation schemes have to be used. In this case no specific assumptions about a model beyond certain regularity of the solution are made.

1.4.1 First one should raise the question of what we expect from the sought representation. We may ask for a representation of a curve that is efficiently codable and storable while we do not care for smoothness and properties depending on it such as e.g. length. In this case the chain encoding of Freeman seems very suitable (Freeman (7) and these proceedings).

1.4.2 A generalization of these concepts to surfaces, in a certain sense, was recently given by B. Schulze (17) at the computing center at Münster. He approximates the surface by another one that is composed of triangles. Hence it is piecewise linear, continuous across the edges of the triangles but not differentiable, in general. The surface is known if the corners of all triangles are specified, this is done in a way that may be viewed as 3-dimensional chain encoding.

2. REPRESENTATION OF FUNCTIONS OF ONE REAL VARIABLE

2.1 Let us first glance at the classical problem of interpolation:

$\mathbb{P}_1$: Given are n data points (x_j, y_j), $j=0,\ldots,n$. Suppose $x_j \neq x_k$

for every pair $j \neq k$. Find a curve that matches the given points and is easy to compute for every x in a specified interval.

2.1.1 The classical solution is to take a polynomial $p(x)$ of degree n that satisfies

$$p(x_j) = a_o + a_1 x_j + \ldots + a_n x_j^n = y_j \quad \text{for} \quad j=0,\ldots,n.$$

This interpolating polynomial is uniquely determined, thus it does not matter how it is constructed, at least theoretically.

Let n be fixed and choose $0 \leq i \leq n$. It is quite simple to find a polynomial $u_{i,n}$ of degree n that is equal to 1 at x_i and zero at all other x_j

$$u_{i,n}(x) = \frac{(x-x_o)\cdot(x-x_1)\ldots(x-x_{i-1})\cdot(x-x_{i+1})\ldots(x-x_n)}{(x_i-x_o)\cdot(x_i-x_1)\ldots(x_i-x_{i-1})\cdot(x_i-x_{i+1})\ldots(x_i-x_n)},$$

of course the first factor in numerator and denominator is missing if i=0, the last one if i=n.

With these "cardinal polynomials" the solution of the general interpolation problem is found as "Lagrange interpolation" formula,

$$p(x) = \sum_{i=o}^{n} y_i u_{i,n}(x) .$$

Another method for the construction of the interpolating polynomial goes back to Newton. Here we write, with previous cardinal polynomials,

$$p(x) = y_o + c_1 u_{1,1}(x) + c_2 u_{2,2}(x) + \ldots + c_n u_{n,n}(x) \ .$$

From $p(x_1) = y_o + c_1 \cdot 1 + c_2 \cdot 0 + \ldots + c_n \cdot 0$ the coefficient c_1 is found. Inserting $x = x_2$ yields the next coefficient c_2, etc. Hence there is a recursive way for the build up of p(x). This principle is generalized to higher dimensions in (26).

2.1.2 One would expect that the interpolating polynomials will converge to the given function f(x) from which the values $y_j = f(x_j)$ are taken if the number of interpolation points is increased. But this depends upon many assumptions compare e.g. Natanson (14) and Rivlin (15) for more details. We only remark that even for some unsuspiciously looking functions and sets of interpolation points convergence need not take place, compare (25).

On the other hand to every continous function f(x) one can find a sequence of interpolating polynomials that will converge to f(x) uniformly. But there is no common rule for finding their interpolation points. Also care is advisable in evaluating high order polynomials because of round-off errors, it is dangerous to "simplify" the calculation by rearranging the terms of the Newton interpolation.

2.1.3 Another method is the interpolation by <u>rational functions</u>

$$R = \frac{a_o + a_1 x + \ldots + a_\ell x^\ell}{b_o + b_1 x + \ldots + b_m x^m} \ , \qquad n = \ell + m \ .$$

Interpolating rationals are not as "stiff" as polynomials but seem to allow for "sharper turns". Their construction is almost as simple as that of an interpolating polynomial and they have proved superior to polynomials in many cases. One can provide a uniform approach to polynomial and rational interpolation (24). But there are even more possible complications than in polynomial interpolation. There is not always a genuine solution for rational interpolation because some points may be "left out" by the curve, (so called unattainable points) and good codes will show this. Even more, the interpolation function $R(x)$ may become infinite between the interpolation points, certainly not a good way of approximating a continuous curve.

Thus rational functions are an excellent tool but should be used judiciously.

It should be noted that interpolation by polynomials and rationals can also handle those data in which the value of the function and its derivatives up to a certain degree are prescribed - the number of data should be n+1.

2.1.4 A fairly new tool are the so called spline-functions, that have been "invented" three decades ago and have been studied systematically in the last years.

A synthetic definition of a spline function may be to say that the domain of definition (an interval $I = [\alpha,\beta]$) is split into subintervals by a number of knots x_j $(x_o = \alpha < x_1 \leq \ldots < x_N = \beta)$. In each of the subintervals $[x_{j-1}, x_j] := \{x \mid x_{j-1} \leq x \leq x_j\}$ the spline is given by a function of, say, n+1 parameters and at the knots the spline has to be n times differentiable, i.e. the right and left limits of the derivatives are equal for $x \to x_j$, any interior knot.

It is know that a polynomial or a rational function is completely determined by the values in any small region, in fact, by its values in n+1 points. The spline relaxes this strong connection. To give an example a spline may be identically zero in one subinterval without being identically equal to zero in the others.

Let us look at the most common type - the cubic splines - piecewise 3rd degree polynomial. Given (x_j, f_j) for $j=0,1,\dots,N$, we first write down the second derivative of a cubic spline. In each subinterval $[x_{j-1}, x_j]$

$$s''(x) = M_{j-1} \frac{x_j - x}{h} + M_j \frac{x - x_{j-1}}{h} \quad \text{with} \quad h = x_j - x_{j-1} ,$$

is a linear function and its values M_j $(j=0,\dots,N)$ at x_j are unknown.

Integrating twice we get with $f_j = \tilde{f}_j - \frac{h^3}{6} M_j$ the formula

$$s(x) = \frac{1}{h}\left[M_{j-1} \frac{(x_j - x)^3}{6} + M_j \frac{(x - x_{j-1})^3}{6} + \tilde{f}_j (x - x_{j-1}) + \tilde{f}_{j-1}(x_j - x) \right] .$$

Using continuity of the first derivatives a system of N-1 linear equations for the M_j is derived, compare (27).

Hence in addition to the interpolation data one may impose two extra conditions. The so called "natural boundary conditions" fix $M_0 = M_N = 0$.

A number of programs for spline interpolations are to be found in Späth (20). In books on applied mathematics the more general cases of higher order polynomials and appropriate boundary conditions are discussed. The mathematically advanced reader is referred to de Boor (3). For the generalization to piecewise rationals and other types of functions see (23).

2.2 APPROXIMATION

2.2.1 To understand, how approximation enters the scene, let us start with a simple example. Suppose that measurements have been made and we have good reasons to believe that they are best approximated by a straight line: $y = a_o + a_1 x$, the coefficients should be determined.

If there are only two points (0,3),(1,6) we find a unique solution:

$$\begin{aligned} a_o + 0 \cdot a_1 &= 3 \\ a_o + 1 \cdot a_1 &= 6 \end{aligned} \quad \Rightarrow \quad a_o = 3,\ a_1 = 3 \quad \text{(line 1 of fig. 1)}.$$

Usually more measurements have been made than necessary in order to improve accuracy, they may have led to (2,5),(3,6) or the equations

$$a_o + 2 \cdot a_1 = 5$$

$$a_o + 3 \cdot a_1 = 6$$

and there is no simultaneous solution to all 4 equations.

What is the best one can do? There are different answers depending upon what is called "best".

One may wish the maximal error to be as small as possible, i.e. if

$$\varepsilon_i = a_o + a_1 \cdot x_i - y_i \ ,$$

"Best approximations" depending upon the specified criterium.

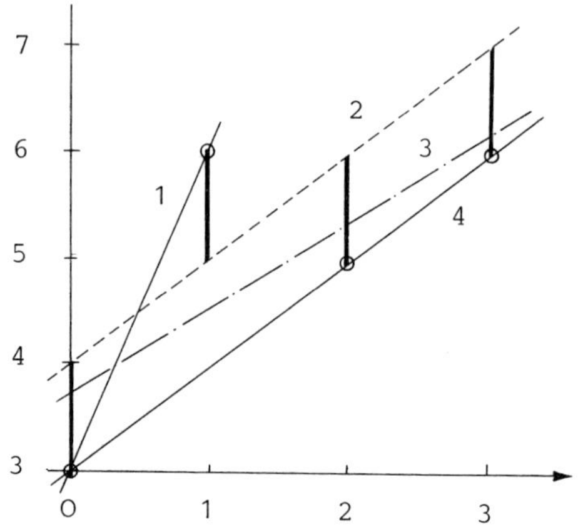

F I G U R E 1

then minimize

$$\max_i |\varepsilon_i| = \|\varepsilon\|$$

by proper choice of a_o and a_1 (<u>discrete Tschebyscheff-Approximation</u>).

The solution is given by $a_o = 4$, $a_1 = 1$

$$\max |4 + x_i - y_i| = 1 = \|\varepsilon\| ,$$

represented by the line 2. It is seen that alternatively the data points lie above and below this line in maximal distance. In other words all above equations are satisfied up to the error $\|\varepsilon\|$. For computational details compare (27). This technique puts high confidence in every data point therefore the worst deviations are made as small as possible.

If the uncertainty of the measurements is taken to be rather large, one should not put too much weight to the largest deviations, they may be due to noise or technical errors. This is done by the <u>least square approximation</u> (Gauß). In our example let us try to minimize

$$F(a_o, a_1) := \sum_{i=o}^{3} \varepsilon_i^2 \qquad (\underline{L^2\text{-Norm}} \text{ of error}) .$$

The solution is found from Gauß' famous <u>normal equations</u>

$$\frac{1}{2}\frac{\partial F}{\partial a_o} = \Sigma(a_o + a_1 x_i - y_i) = 4a_o + 6a_1 - 20 = 0$$

$$\frac{1}{2}\frac{\partial F}{\partial a_1} = \Sigma(a_o + a_1 x_i - y_i)x_i = 6a_o + 14a_1 - 34 = 0$$

with the solution $a_o = \frac{19}{5}$, $a_1 = \frac{4}{5}$, plotted in line 3. While we have $F(4,1) = 4$ in case of the Tschebyscheff solution we obtain $F(3.8, 0.8) = 2.8$ for the Gauß-approximation.

A not too serious remark: A careful control of the measurements might have revealed that the second point was completely wrong because the apparatus did not function properly and the 3 other points were fitted ideally by a straight line 4 giving <u>the</u> answer.

This example may look oversimplified but it illustrates all features of the mathematical method.

2.2.2 If a function $f(x)$ in I shall be stored in a computer and this is done by means of rationals or splines then the difference between the approximation $R(x)$ and the function $f(x)$ shall be <u>uniformly</u> small:

$$|f(x) - R(x)| \leq \varepsilon \quad \text{for every } x \text{ in the interval I.}$$

If one asks for that R(x) which minimizes max|f(x) - R(x)|, the resulting function is called the Tschebyscheff-Approximation of f(x). May be the function is known at a discrete point set X of m points x_j (j=1,...,m) only and we have good reasons to assume that it is essentially fitted by a polynomial of fixed degree n, for instance a straight line, then we could use methods as described in (27) to find those coefficients of the polynomial, for which $\max_X |f(x) - P(x)|$ is smallest. The method (Remes' algorithm) is closely related to the linear optimization, but specialized for this case and dating back to the 30's. This kind approximation is also called (discrete) Tschebyscheff-Approximation. Its use can be recommended only if the input data $f(x_j)$ are known to be (almost) exact.

2.2.3 If the values $f(x_j)$ arise from measurements it is much more advisable to resort again to the Gaussian method of least square fitting. For discrete data the expression

$$\sum_{j=1}^{M} (f(x_j) - R(x_j))^2 = F(a)$$

and for continuous functions (practically only a fictitious situation) the mean square deviation

$$\int_{\alpha}^{\beta} (f(x) - R(x))^2 dx = F(a)$$

is minimized, where a is to denote the parameters that determine the function R(x), (1, 4).

2.2.4 An interesting result is obtained for least square approximation by polynomial splines with respect to its derivatives.

For the given points (x_j, f_j), $j=0,\ldots,N$ of a function $f(x)$ we may ask for that cubic spline $s(x)$ with the knots x_j, for which the mean square deviation of the second derivatives s'' from that of the function $f(x)$ is smallest, so that $s(x)$ is smoothing $f(x)$ in some special sense. The answer is given by that cubic spline which interpolates at the given knots and the first derivatives of which at the interval endpoints agree with $f'(\alpha)$ and $f'(\beta)$, respectively.

The natural splines (resulting from $f'(\alpha) = f'(\beta) = 0$) are characterized by the property that they minimize $\int |f''|^2 dt$ among all functions interpolating f at the knots.

2.3 Sometimes one likes the function interpolating or approximating given data to retain certain typical properties like monotonicity or convexity.

It is easily seen from e.g. the cardinal polynomial $u_{6,6}(x)$ in 2.1.1 that polynomials are not well suited for this purpose.

2.3.1 In case of convexity one could use special rational splines (Schaback, Diss. 1969/Münster) as described in (23) to solve the interpolation problem and retain convexity.

As a further example we quote a result by Hornung (10). He shows that one can find a convex cubic spline satisfying the interpolation conditions that has, however, possibly $\frac{3}{2}N$ knots, some of them may be the given points, but others of the given x_j may have to be split into two "neighboring" knots.

3. INTERPOLATION OF FUNCTIONS OF TWO REAL VARIABLES

3.1 The set up of the interpolation problem for two (or more) independent variables is literally the same as before:

$\mathbb{P}_2$: Given are N data points $((x_j,y_j),f_j)$ $j=1,\dots,N$ with $(x_j,y_j) \neq (x_k,y_k)$, that is to say not both coordinates are equal, and we ask for a surface that will match these points.

3.1.1 Again one may turn to polynomials for the solution but in this case it is not even clear, what class of polynomials to use. Typical alternatives are

$$\mathcal{P}_{m,n} = \left\{p(x,y) \mid p(x,y) = \sum_{i=0}^{m} \sum_{k=0}^{n} a_{ik}\, x^i\, y^k\right\}$$

$$\mathcal{P}_{n} = \left\{p(x,y) \mid p(x,y) = \sum_{i+k\leq n} a_{ik}\, x^i\, y^k\right\}$$

which are apparently different.

It is obvious that the number of coefficients, which is easily found from m and n resp. n, has to be equal to the number N of given data for the problem to possess a solution. Then one could set up linear equations

$$\Sigma a_{ik}\, x_j^i\, y_j^k = f_j \qquad (j=1,\dots,N)$$

to determine the coefficients.

It is obvious that one cannot expect a solution in every case. If, e.g., the points (x_j,y_j) all lie on a straight line, say the x-axis, and there are more than two points (i.e. $N > 2$) the coefficients will not be uniquely determined or the system will have no solution at all, if $\mathcal{P}_{m,n}$ is used and $n > 0$.

The conclusion can be drawn that in the two (and higher) dimensional interpolation problem there is no affirmative answer to $\mathbb{P}_2$ but additional assumptions are necessary.

3.1.2 One will therefore turn attention to well structured sets of interpolation points. A typical case is a set forming a rectangular grid.

To describe this set denote by

$x_o,\ldots,x_m$ the x-coordinates

$y_o,\ldots,y_n$ the y-coordinates

and every (x_j,y_ℓ) for $j \in \{0,\ldots,m\}$ and $\ell \in \{0,\ldots,n\}$ is interpolation point. For simplicity we may denote the associated given interpolation data value by f_j . In this case the interpolation polynomial of class $\mathcal{P}_{m,n}$ is easily given explicitly. The construction proceeds along the following line.

1) Find polynomials $w_{j,\ell}(x,y)$ which are 1 at (x_j,y_ℓ), one fixed grid point, and attain the value zero at every other grid point. (<u>Cardinal polynomials</u>).

2) Having obtained the cardinal polynomials the interpolation problem is solved by

$$p(x,y) = \sum_{i=0}^{m} \sum_{k=0}^{n} f_{ik} w_{ik}(x,y) \in \mathcal{P}_{m,n} .$$

This solution is unique. Of course it would mean a big

reordering to find the originally asked for coefficients a_{ik} and it is advised not to do this unless really these coefficients are needed (say for theoretical purpose).

To construct the $w_{j,\ell}(x,y)$ one may go back to 2.1.1, to find polynomials of degree m in x and that satisfy

$$u_j(x_k) = \begin{cases} 0 & \text{if } k \neq j \\ 1 & \text{if } k = j \end{cases},$$

similarly n-th degree polynomials in y with

$$v_j(y_k) = \begin{cases} 0 & \text{if } k \neq j \\ 1 & \text{if } k = j \end{cases}.$$

It is now immediately seen by inserting the grid points as arguments that

$$w_{j,\ell}(x,y) = u_j(x) \cdot v_\ell(y)$$

is the required two dimensional cardinal polynomial.

Some people call the set $\{w_{j,\ell}(x,y)\}$, $j=0,\dots,m$, $\ell=0,\dots,n$, the <u>tensor product</u> of the sets $\{u_j(x)\}$ and $\{v_\ell(y)\}$.

A generalization of these concepts to the case that not every grid point is used for interpolation is found in (26). One may use the Newton type interpolation for calculation if the interpolation points are ordered properly.

3.1.3 As a side remark we may look at the so called <u>blending interpolation</u>. Using the notations of the previous section and assuming that $f(x,y)$ can be evaluated at points $f(x_k,y)$ and $f(x,y_\ell)$ (e.g. by first evaluating $f(x_k,y)$ for a large number of y and replacing $f(x_k,y)$ by an approximation in the one variable y as described in chapter 2) one may form the function

$$g(x,y) = \sum_j u_j(x)\cdot f(x_j,y) + \sum_k v_k(y)\cdot f(x,y_k) - \sum_{j,k} f(x_j,y_k)\cdot w_{j,k}(x,y) \ .$$

It is interesting to look for $g(x_\ell,y)$, x_ℓ a given x-coordinate of a grid point. Observing $u_\ell(x_\ell) = 1$ and $u_j(x_\ell) = 0$ if $j \neq \ell$, one obtains from the definition, since all summations over j can be evaluated:

$$g(x_\ell,y) = f(x_\ell,y) + \sum_k v_k(y)\cdot f(x_\ell,y_k) - \sum_k f(x_\ell,y_k)\cdot v_k(y) = f(x_\ell,y) \ .$$

The functions g and f are equal along the lines $x = x_\ell$.

In analogous way one verifies

$$g(x,y_\ell) = f(x,y_\ell) \ .$$

The function $g(x,y)$ is not a polynomial, but it forms an approximation of $f(x,y)$ (with its two independent variables) from functions $f(x_j,y)$ resp. $f(x,y_k)$ (of only one variable) by combining them with polynomial coefficients.

Naturally blending interpolation gives a much better approximation of $f(x,y)$ than the tensor interpolation explained before which pins the interpolating polynomial to the given function only at the grid points.

3.1.4 It is obvious that instead of the cardinal polynomials one could use cardinal splines to obtain two dimensional spline interpolation on data given at a rectangular grid. Very detailed explanations and references can be found in (16), e.g. the papers by Hall (8).

3.1.5 We briefly point out that there are many interesting applications of the type of polynomial interpolation given in (26) in the field of <u>finite elements</u>. In this case the approximations of a function f(x,y) are pieced together by appropriate approximations, frequently polynomials, defined on "cells" into which the domain of f(x,y) is subdivided. Favourite cells are rectangles and triangles. As a guide to the literature compare Whiteman (28).

3.1.6 <u>Interpolation on triangles</u>. Let $P_j = (x_j, y_j)$, j=1,2,3 denote the corners of the triangle. The area of the triangle, by elementary analytic geometry, is given by

$$F(P_1,P_2,P_3) := \frac{1}{2} \det \begin{vmatrix} x_1 & y_1 & 1 \\ x_2 & y_2 & 1 \\ x_3 & y_3 & 1 \end{vmatrix} \quad \text{(apart from its sign).}$$

If P = (x,y) is any point on the straight line through P_2 and P_3, then $F(P,P_2,P_3) = 0$. Hence

$$w_1(x,y) = \frac{F(P,P_2,P_3)}{F(P_1,P_2,P_3)}, \quad w_2(x,y) = \frac{F(P_1,P,P_3)}{F(P_1,P_2,P_3)}, \quad w_3(x,y) = \frac{F(P_1,P_2,P)}{F(P_1,P_2,P_3)}$$

are <u>cardinal linear functions</u> for the given points,

$$w_i(P_j) = \begin{cases} 1 & j = i \\ 0 & j \neq i \end{cases} .$$

The linear function matching data f_i at P_i is given by

$$p(x,y) = \sum_{j=1}^{3} f_j w_j(x,y) \in \mathcal{P}_1 .$$

Functions which attain given first derivatives in addition to the prescribed function values may be obtained as the sum of terms

$$[a_i + b_i(x-x_i) + c_i(y-y_i)] \cdot w_i^2(x,y) , \quad i=1,2,3 .$$

3.2 APPROXIMATION

In case of several variables again one may formally procede just as in the case of one independent real variable and ask for a fit which is good with respect to one or the other criterion. Usually the square sum (or integral) or the maximum of the absolute value of the error is made as small as possible.

3.2.1 Let D be a domain in which the data points $P_j = (x_j, y_j)$, $j=1,\ldots,N$ are located. Then one may approximate by expressions $H(x,y;a)$ defined for (x,y) of D and a from an n-dimensional parameter set.

If given data (P_j, f_j) are to be approximated in L^2 sense, one has to minimize

$$g(a) := \sum_{j=1}^{N} |H(x_j, y_j; a) - f_j|^2 ,$$

which leads to

$$\frac{\partial g}{\partial a_i} = \sum_{j=1}^{N} (H(x_j, y_j; a) - f_j) \cdot \frac{\partial H}{\partial a_i}(x_j, y_j; a) = 0$$

and these are again equations for the unknown coefficients $a = (a_1, \ldots, a_n)$. We refer to (4) and books of numerical algebra for solution techniques. This method is useful in connection with smoothing of noisy data (say measurements).

3.2.2 The minimization of

$$g(a) = \max_j |H(x_j, y_j; a) - f_j|$$

by proper choice of a (i.e. Tschebyscheff-approximation) can be reduced to a problem of optimization. In fact the program can be specially adapted to this case and will be a generalized form of the Remes-algorithm. The result is an approximation with a guaranteed maximal bound for the deviation from the given data values f_j and this is reasonable only if these values are (rather) precise. In both cases (L^2-or Tscheby-scheff-Approximation) the numerical work may be fairly large if N is large.

3.2.3 If the data points are scattered considerable pain has been taken to find numerically economic methods that can compete with all out best approximation. Schumaker (18, 19) gave a survey of ideas up to 1976 that have been invented for that purpose.

3.2.4 In some applications it may be important to fit measured data by functions that preserve the area resp. volume between the function and (say) the line resp. plane $f \equiv 0$. These techniques are useful in replacing a histogram by a smooth function. Using the techniques developed by Kendall and coworkers (2) programs for the two dimensional case were made available by Herden (9). We refer the reader to (25).

3.2.5 This brings up the question of defining splines for general distribution of data points. Recently several attempts have been made in this direction. Starting from the abstract concept of the spline definition in one variable multivariate versions were proposed by Laurent (11) and Meinguet (12). Another method that starts from algebraic properties of the spline concept is to appear in a paper by Micchelli (13).

REFERENCES

1. Bestehorn, M., Ebert, U., Steinhausen, D., Werner, H.: Some Applications of Computational Mathematics to Medical Problems, in "Computing Methods in Applied Sciences and Engineering, 1977, I" (R. Glowinski and R.J. Lions, ed.), Springer Verlag, Berlin, Heidelberg, New York, (1979).

2. Boneva, L.I., Kendall, D., Stefanov, I.: Spline-Transformations: Three New Diagnostic Aids for the Statistical Data-Analyst, J. of Royal Stat. Soc., Ser. B 33 (1971).

3. de Boor, C.: A Practical Guide to Splines, Applied Mathematical Sciences 27, Springer, New York, Heidelberg, Berlin, (1978).

4. Braess, D.: Nichtlineare Anpassung von Meßdaten nach der Methode der kleinsten Quadrate, Schriftenreihe des Rechenzentrums der Universität Münster, Nr. 6 (1974).

5. Delvos, F.J., Posdorf, H.: N-th Order Blending, in "Constructive Theory of Functions of Several Variables" (W. Schempp, K. Zeller, ed.), Springer, Lecture Notes in Math. 571, (1977).

6. Delvos, F.J., Posdorf, H.: A Representation Formula for Reduced Hermite Interpolation, in "Numerische Methoden der Approximationstheorie, Bd. 4" (L. Collatz, G. Meinardus, H. Werner, ed.), ISNM Vol. 42, Birkhäuser, Basel, Stuttgart, (1978).

7. Freeman, H.: Computer Processing of Line-Drawing Images, Computing Surveys 6 (1974).

8. Hall, C.A.: Spline Blended Approximation of Multivariate Functions, in Sahney (16).

9. Herden, W.: Darstellung von Histogrammen durch Splinefunktionen, Staatsexamens- und Diplomarbeit Münster, (1978).

10. Hornung, U.: Interpolation by Smooth Functions under Restrictions for the Derivatives, Schriftenreihe des Rechenzentrums der Universität Münster, Nr. 33 (1978).

11. Laurent, J.P.: Algorithms for Two-Dimensional Splines, in "Approximation in Theorie und Praxis" (G. Meinardus, ed.) Bibliographisches Institut, Mannheim.

12. Meinguet, J.: An Intrinsic Approach to Multivariate Spline Interpolation at Arbitrary Points, in Sahney (16).

13. Micchelli, Ch.: A Constructive Approach to Kergin-Interpolation in $\mathbb{R}^k$: Multivariate B-Splines and Lagrange Interpolation, Rocky Mountain J. of Math. to appear

14. Natanson, I.P.: Theorie der Funktionen einer reellen Veränderlichen, Akademie-Verlag, Berlin, (1961).

15. Rivlin, Th.: The Chebyshev Polynomials, John Wiley & Sons, New York, London, Sydney, Toronto, (1974).

16. Sahney, B.N.: Polynomial and Spline Approximation, Reidel, Dordrecht, Boston, London, (1979).

17. Schulze, B.: Ein Code zur Approximation von Hyperflächen im $\mathbb{R}^n$, Schriftenreihe des Rechenzentrums der Universität Münster, Nr. 37 (1979).

18. Schumaker, L.L.: Fitting Surfaces to Scattered Data, in "Approximation Theory II" (G.G. Lorentz, C.K. Chui, L.L. Schumaker, ed.), p. 203-268, Akademic Press, New York, San Francisco, London, (1976).

19. Schumaker, L.L.: Two-Stage Spline Methods for Fitting Surfaces, in "Approximation Theory" (R. Schaback and K. Scherer, ed.), Springer, Lecture Notes in Math. 556, (1976).

20. Späth, H.: Spline-Algorithmen zur Konstruktion glatter Kurven und Flächen, Oldenbourg, München, Wien, (1973).

21. Späth, H.: Cluster-Analyse-Algorithmen zur Objektklassifizierung und Datenreduktion, Oldenbourg, München, Wien, (1975).

22. Steinhausen, D., Langer, K.: Clusteranalyse: Einführung in Methoden und Verfahren der automatischen Klassifikation, Walter de Gruyter, Berlin, New York, (1977).

23. Werner, H.: An Introduction to Non-Linear-Splines, in Sahney (16).

24. Werner, H.: Ein Algorithmus zur rationalen Interpolation in "Numerische Methoden der Approximationstheorie, Bd. 5" (L. Collatz, G. Meinardus, H. Werner, ed.), ISNM Vol. 52, Birkhäuser, Basel, Stuttgart, (1979).

25. Werner, H.: Numerical Algorithms for Interpolation and Smoothing, Schriftenreihe des Rechenzentrums der Universität Münster, Nr. 42, (1979).

26. Werner, H.: Remarks on Newton type multivariate Interpolation for Subsets of Grids, Computing to appear.

27. Werner, H., Schaback, R.: Praktische Mathematik II, 2. Auflage, Berlin, Heidelberg, New York, Springer, (1979).

28. Whiteman, J.R.: A Bibliography for Finite Elements, Academic Press, London, New York, San Francisco, (1975).

COMPUTER GENERATION OF SHADED RELIEF MAPS

Bruce Schachter

Ground Systems Department
General Electric Company
Daytona Beach, Florida

Computer generated shaded relief maps have been used in cartography for a number of years. Most current techniques produce monochromatic images which are simple orthographic projections of the terrain surface. This paper shows how the same data can be used to create more realistic displays. The images generated are full colored, textured, terrain scenes.

I. INTRODUCTION

Topographic maps provide an efficient and unique means of communicating geographical information. They are useful in many activities including the appraisal and development of natural resources, regional planning, the analysis of line-of-site communication systems, and as aids to navigation by sea and air. Topographic information is required for the planning and execution of military operations, defensive and offensive, strategical and tactical.

A topographic map should clearly depict variations in the Earth's slope and elevation. This involves representing a third dimension on a two-dimensional surface. Elevation is usually displayed only symbolically, such as by contour lines. However, map makers occasionally represent variations in surface height by what they call "hill shading". Hill shading

ISBN 0-12-267180-5

makes use of the optical theory which states that the brightness of any small area of a perfectly diffuse undulating surface varies as the cosine of the angle of incident parallel light.

In recent years, the computer has been used to create more accurate shaded relief maps than can be drawn by hand. Certain projections of these maps may be constructed by using the principles of perspective. Most current systems produce gray level images which are simple orthographic projections of a terrain surface. An orthographic projection puts the perspective center at infinity. In this paper, we will provide a technique for creating more realistic displays. The images which we produce are full color perspective projections of a terrain surface. We will show how texture can be added to a scene to enhance realism.

II. U.S. DEFENSE MAPPING AGENCY DATA

The U.S. Defense Mapping Agency (DMA) has assembled digital map data for certain areas of the Earth. Their source material includes aerial photographs, topographic charts, remotely sensed imagery, and ground truth information. DMA data is intended for such applications as static map generation, radar land mass simulation, flight simulation, and cockpit displays for high-speed, low-flying aircraft.

The DMA supplies two categories of data (1): terrain and "culture" files. A <u>terrain file</u> is an array of elevation values for a region of the Earth. The sampling interval of the terrain is a function of latitude (Table I). The elevation array for each terrain region is stored on magnetic tape at two different resolutions.

TABLE I. DMA Terrain Data Intervals

	Level 1	Level 2
Latitude	Lat. × Lon.	Lat. × Lon.
0°-50° N/S	3 × 3 seconds	1 × 1 second
50°-70° N/S	3 × 6 seconds	1 × 2 seconds
70°-75° N/S	3 × 9 seconds	1 × 3 seconds
75°-80° N/S	3 × 12 seconds	1 × 4 seconds
80°-90° N/S	3 × 18 seconds	1 × 6 seconds

A culture file holds the descriptions of manmade and ecological surface features within a given terrain region. Most feature types are represented by a polygonal boundary and a coded description. However, the locations of some features such as bridges, dams, walls, and pipelines are specified by polygonal lines; while others such as tall buildings and water towers are given as points.

Culture polygons are assigned both a predominant surface material and a more specific surface description. The 13 general categories of surface materials are given in Table II. As an example, the general category assigned to a culture polygon may be "trees". The particular subclasses under this heading are orchard, deciduous forest, coniferous forest, evergreen, and mixed forest. Each tract of trees is assigned a predominant tree height

TABLE II. Thirteen DMA Predominant Material Types

1. Metal	8. Rock
2. Part Metal	9. Asphalt/Concrete
3. Stone/Brick	10. Soil
4. Composition	11. Marsh
5. Earthen Works	12. Trees
6. Water	13. Snow/Ice
7. Desert/Sand	

III. EQUATIONS FOR DISPLAYING A SHADED RELIEF MAP

Suppose that a terrain surface H(x,y) is illuminated by both direct sunlight and ambient light. The location of the sun is given in terms of its azimuth θ and elevation ϕ. Azimuth is measured clockwise from North. Elevation is the angle between the sun and the horizon (Figure 1).

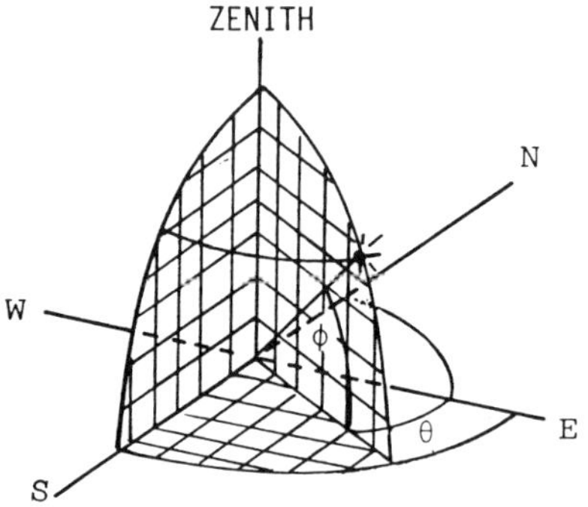

FIGURE 1. ϕ = Angular Elevation of the Sun; θ = Sun Azimuth

Let P_λ denote the quantity of direct sunlight of frequency λ reaching the terrain surface. Under a simple model, the amount of sunlight reflected by a surface element is related to the cosine of the incident angle γ. Let $\vec{n}(x,y)$ denote the unit normal at surface element (x,y) and let $\vec{n}_p$ be a unit vector in the direction of the sun (Figure 2). Then

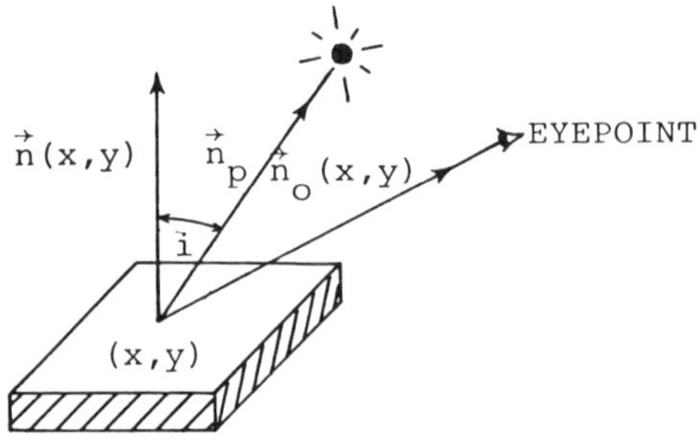

FIGURE 2. Unit Direction Vectors

$$P_\lambda \cos\gamma = P_\lambda [\vec{n}(x,y) \cdot \vec{n}_p]$$

where

$$\vec{n} = \left[\frac{-\partial H(x,y)}{c\partial x}, \frac{-\partial H(x,y)}{c\partial y}, 1/c\right],$$

$$c = \left\{1 + \left[\frac{\partial H(x,y)}{\partial x}\right]^2 + \left[\frac{\partial H(x,y)}{\partial y}\right]^2\right\}^{1/2}$$

and

$$\vec{n}_p = [\sin\theta \cos\phi, \cos\theta \cos\phi, \sin\phi].$$

We will also assume that the surface is illuminated by ambient light. Ambient light strikes a surface element equally from all directions and is assumed to be reflected equally in all directions. The intensity of ambient light for wavelength λ will be denoted by A_λ.

The albedo of an object is the fraction of the incident energy which is reflected by the object over the entire electromagnetic bandwidth. The albedo of a surface for frequency λ will be denoted by r_λ. It is the fraction of the energy of that wavelength which is reflected. Tables exist for r_λ for many materials (7-11). The visual albedo of an object refers to the object's albedo as perceived by the human visual system. It is the product of the object's albedo and the spectral luminosity of the human eye.

A number of simplifying assumptions are needed before we can present an equation for the appearance of a surface. We will assume that reflected light from one surface element does not illuminate another surface element. This effect is difficult to model and will lead to global computations.

We will assume that surface elements do not cast shadows on each other. This restriction can be lifted by using one of the shadow algorithms given in the computer graphics literature.

Computer generated images may be viewed on a color cathode-ray tube (CRT) or transferred to film. It is important that the images generated accurately represent the calculated computer model. The spectral emission characteristics of a typical color monitor can produce a fairly accurate rendition of a color scene. However, the color characteristics are far from ideal for photographic recording purposes. The equations presented in this paper do not include compensating terms for the spatial nonlinearities of a CRT or its emission characteristics. Detailed discussions of these problems may be found in (3), (5), (6), and (14). Photographs taken by a normal camera off of a color video display are in general of poor quality. Much better results may be obtained by using one of the film recorders specially designed for this purpose; such as those manufactured by Dunn Instruments Incorporated, Matrix Corporation, or Dicomed Corporation.

Under the above assumptions, the appearance of surface element (x,y) over the visual bandwidth is given by

$$[R_\lambda(x,y) \mid 380 \text{ nm} < \lambda < 770 \text{ nm}],$$

where

$$R_\lambda(x,y) = P_\lambda \max\,[0, \vec{n}(x,y) \cdot \vec{n}_p]\; r_\lambda(x,y) + A_\lambda r_\lambda(x,y).$$

Before an image is displayed on a color cathode-ray tube, frequency values are usually converted first to CIE tristimulus values and then to red, green and blue intensity levels.

These transformations are described in detail in (5) and (14). The color image R(x,y) may be called the Reflectance Map of the field.

An example of a DMA terrain file is shown in Figure 3a. Here, high gray levels are used to represent high elevations. The corresponding culture map is given in Figure 3b. A "sun-shaded" overhead view of terrain with culture is displayed in Figure 3c. A perspective projection of the scene is shown in Figure 3d. It was created by using the algorithm given in the Appendix.

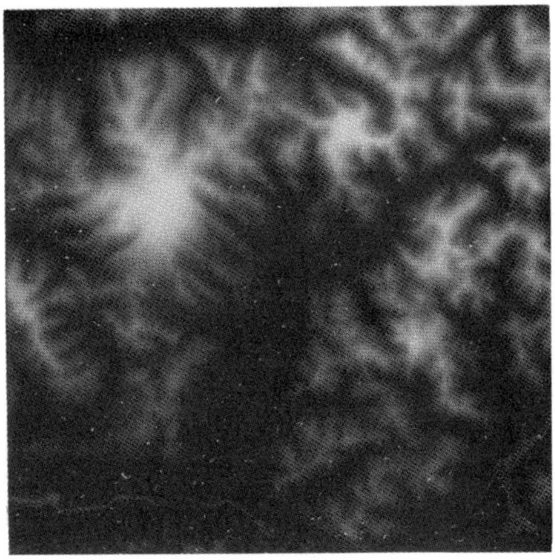

Figure 3a. Terrain Map, High Gray Levels Represent High Elevation Values

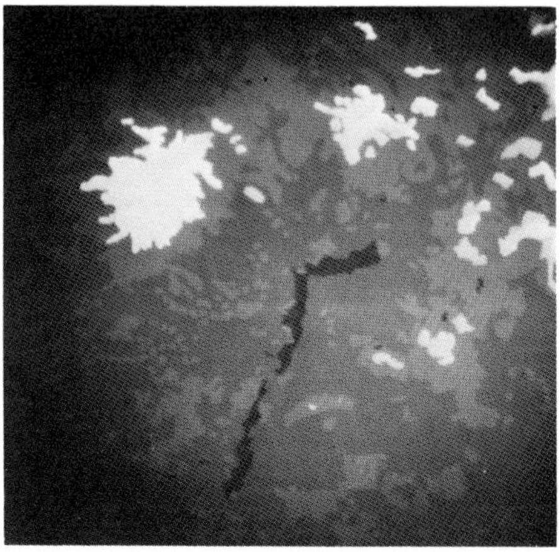

Figure 3b. Culture Map, (Note: All Photographs are Black and White Versions of Original Color Photographs)

Figure 3c. Sun-Shaded Terrain With Culture

Figure 3d. Terrain with Culture, In Perspective

IV. ADDING TEXTURE TO A SCENE

When we view a terrestrial landscape from the air, we see a mosaic of different plant species and land uses. Expensive agricultural machinery forces farmers to plant their crops in large rectangular fields. Cities, forests, orchards, swamps, oceans, and deserts all have distinctive appearances. The individual characters of some of these surface features are due mainly to their color. However, others are distinguished by both color and texture. We may sample aerial photographs to obtain prototype textures. The area scanned should be flat, fairly homogeneous, and free of ecological anomolies. These prototype patterns will be put onto a terrain surface in the following manner.

First, we use DMA elevation data to create an overhead view of a shaded relief map. Then we identify a polygonal region on the map which is to receive a particular texture. The appropriate texture prototype is then used to modulate the color intensity of this region.

It should be understood that this process involves an inherent distortion of the texture array, since we are projecting a planar pattern onto a three-dimensional surface. Also, the sun angle and elevation of the texture prototype may not correspond to that of the image which we are creating. Nevertheless, texture very effectively adds to the realism of a scene. A future project will replace these texture prototypes by artificially generated textures. This will give us more control over the texture's characteristics (12).

Figure 4a shows a prototype forest texture. Figure 4b was produced by modulating the color within forest regions of Figure 3c. A perspective projection is shown in Figure 4c.

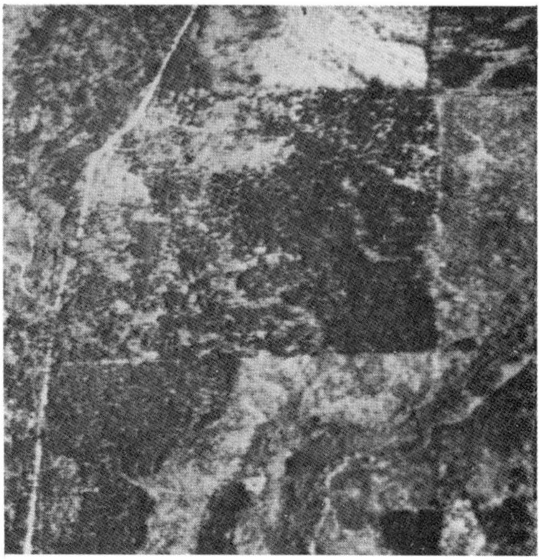

Figure 4a. Texture Array
(Deciduous Forest)

Figure 4b. Terrain with Culture
and Texture

Figure 4c. Terrain with Culture and Texture, In Perspective

ACKNOWLEDGMENT

I would like to thank Barry Fishman for his help in preparing this paper.

REFERENCES

1. ---, Product Specification for Digital Landmass Systems Data Base, Defense Mapping Agency, St. Louis AFS, Missouri, July 1977.

2. W. Bunker and R. Heartz, Perspective Display Simulation of Terrain, Air Force Human Resources Lab., AFHRL-TR-76-39, June 1976.

3. E. Catmull, A tutorial on compensation tables, in Computer Graphics: A Quarterly Report of SIGGRAPH-ACM 13(2), Ed. B.W. Pollack, 1979, 1-7.

4. W. Dungan, A terrain and cloud computer image generation model, in Computer Graphics: A Quarterly Report of SIGGRAPH-ACM 13(2), Ed. B.W. Pollack, 1979, 143-146.

5. R.F. Ferreland and J.M. Booth, Design Handbook for Imagery Interpretation Equipment, Boeing Aerospace, Seattle, Washington, 1975.

6. L.S. Gruber and H. Barney, Color computer graphics and imaging with Polaroid 8×10 POLACOLOR Land Film and the Dunn 631 color camera, Harvard Computer Graphics Conference, Cambridge, Mass., July 1979.

7. C.D. Hendley and S. Hect, The colors of natural terrain and their relation to visual color deficiencies, J. Optical Society of America 39(10), 1949, 870-873.

8. E.L. Krinov, Spektral'naia Otrazhatel'nia Sposbnost' Prirodynykh Obrazovanii (Special Reflectance Properties of Natural Formations), Lab. Aerometodov, Akad, Nauk SSSP, Moscow, 1947.

9. D. Nickerson, K.L. Kelly, and K.F. Stultz, Colors of soils, J. Optical Society of America, 1945, 297-300.

10. R. Penndorff, Luminous and Spectral Reflectances as well as Colors of Natural Objects, Air Force Cambridge Research Center, Bedford, Mass., (Geographical Research Paper 44, 9ASTIZ No. AD-98766, AFCRS-TR-56-203), 1956.

11. R.L. Pendleton and D. Nickerson, Soil colors and special Munsell color charts, Soil Science 71, 1951, 35-43.

12. B. Schachter and N. Ahuja, Random pattern generation processes, Computer Graphics and Image Processing 10, June 1979, 95-114.

13. B.J. Schachter and B. Fishman, Computer Display of Height Fields, General Electric Tech. Pub., Daytona Beach, Florida, 1979.

14. A.P. Verdi, Reproduction of Color in Computer Image Generation Displays, General Electric Tech. Pub., Daytona Beach, Florida, 1979. (Not available for distribution)

15. R.H. Wallis, Film Recording of Digital Color Images, Univ. of Southern California Image Processing Institute, Report 570, Los Angeles, California, 1975.

APPENDIX

ALGORITHM FOR VIEWING A HEIGHT FIELD IN PERSPECTIVE

Provide the following parameters. Typical values are shown in parenthesis.

FOV = field of view (.6)

VSC = vertical scaling constant (.3)

DOF = depth of field (1000)

ALPHA = view angle counterclockwise from x-axis ($\pi/2$)

V = (VX,VY,VZ) = viewpoint (256,0,200)

IH = line number of horizon

H(X,Y) = height field

R(X,Y) = reflectance map

NOTE: A screen size of 512-by-512 is assumed.

```
     DO LINE = 0,511
          DO I = 1,512
          BUFF(I) = sky color
          END DO
     TANG = (LINE-256)*FOV/256
     GAMMA = ARCTAN(TANG)
     DX = COS(ALPHA-GAMMA); DY + SIN(ALPHA-GAMMA)
```

```
     MCOL = 0
     RANGE = 0
     Y = VY; X = VX
A:        IF (RANGE .GT. DOF .OR. MCOL .GT. 511 GO TO C
     T = VSC*H(X,Y) - VZ
          IF (RANGE .LT. .01)
          TANB = SIGN(.3*FOV,T)
          ELSE
          TANB = T/RANGE
          END IF
     ICOL = MIN[(512-IH)+TANB*256./FOV,511]
          IF (ICOL .LT. MCOL) GO TO B
          IF (MCOL .EQ. 0) MCOL = ICOL
          DO K = MCOL,ICOL
          BUFFER(K) = R(X,Y)
          END DO
     MCOL = ICOL + 1
B:   X = X + DX; Y = Y + DY
     RANGE = RANGE + COS(GAMMA)
     GO TO A
C:   Scan BUFFER out on CRT row LINE
     END DO
```

INDEX

N

P

Q

R

S

T

U

V

W

Y

Z